Vacuum Technology

Practical Heat Treating and Brazing

Edited by
Roger Fabian

Acquisitions/Editorial
Veronica Flint

Production Project Manager
Suzanne Hampson

Production/Design
Dawn Levicki

First printing, July 1993
Second printing, February 1995
Third printing, November 1998

Library of Congress Cataloging Card Number: 93-71050
ISBN: 0-87170-477-3

ASM International®
Materials Park, OH 44073-0002

Printed in the United States of America

About the Editor . . . . .

ROGER J. FABIAN is Vice President, Eastern Region Manager of Lindberg Heat Treating Company which has 22 operating divisions in the United States. Mr. Fabian has both a BME and an MBA from Rensselaer Polytechnic Institute and has spent his entire career with Lindberg Heat Treating, joining their Boston Division as Plant Metallurgist in 1962, and transferring to the newly built Berlin Division in 1967 as Chief Metallurgist and Quality Control Manager. In 1979 he was promoted to Division Manager, Berlin CT, and in 1985 became Vice President and Eastern Region Manager.

Made a Fellow of ASM International in 1992, currently he is Vice Chairman of the ASM Technical Division Board, a member of the ASH Heat Treating Steel Panel and Chairman of the Immediate Needs Sub Committee; past Chairman of the ASM Heat Treat Council and Chairman of the Editorial Policy Committee for The Journal of Heat Treating; and Past Chairman of the Hartford Chapter of ASM. Also active in other societies, he is a member of the Metal Treating Institute Board of Trustees, Metal Treating Institute Regulations and Safety Committee, and Metal Treating Institute Quality Assurance/Military Specifications Committee, and Past Chairman of the Atlantic Coast Chapter, Metal Treating Institute.

Mr. Fabian has participated in ASM National Conferences, speaking on Vacuum Heat Treating, Vacuum Brazing, and all areas of Heat Treating and Metals Joining. He has also participated in Technical Sessions at the local chapter level.

About the Authors . . . . .

SHELDON KENNEDY trained as a mechanical engineer, and devoted his career to the development, design and sales of specialized automatic production equipment: primarily that used for vacuum processing. Beginning in the mid-1940's with the development of TV tube producing equipment for GE, he has built or specified and sold vacuum processing equipment for National Research Corporation and Vacuum Industries, among others.

At Vacuum Industries, he was Product Manager for sintering furnaces and responsible for the hard metal, metal injection molded parts, and engineering ceramics sintering furnace lines. Mr. Kennedy recently retired from Vacuum Industries.

JOHN GRIFFIN has a degree in Engineering from Northeastern University and over 35 years industry experience. Mr. Griffin is Director of Engineering for Abar Ipsen, Vacuum Industries, and Centorr Furnaces and is responsible for engineered products for these companies.

CARMEN PAPONETTI has a background in Business Management and received a diploma in Metallurgy from the Applied Technology Institute. He has 25 years of "hands-on" experience in the brazing and heat treat fields, with diverse experience in advanced and experimental technology, and special emphasis in continuous brazing and heat treating in hydrogen, exothermic and mixed atmospheres, together with vacuum brazing and heat treating of all alloys. Mr. Paponetti is Vice President of the Metal Joining Group of HI TecMetal Group, a Cleveland, Ohio based company with 12 operating divisions in Northeast Ohio.

ROBERT F. GUNOW, JR., educated in the United States and Canada, is a member of ASM International, American Welding Society, Engineering Society of Detroit, and is a Trustee of the Metal Treating Institute. Active in brazing, heat treating and other thermal processing for the aerospace, medical, nuclear, automotive and chemical industries, Mr. Gunow is President of Vac-Met, Inc.

ALEXANDER J. GUNOW obtained a Ph.D., M.S. and B.A. from the University of Colorado. He serves currently as a technical representative for Vacuum Furnace Systems (VFS) and Therm Alliance Co. His technical activity involves applications regarding high-pressure gas quenching, brazing, heat treating, and other thermal processes. Dr. Gunow is President of Integrated Technologies.

JAMES G. CONYBEAR, a member of ASM International, SAE and AIME, is a graduate in Metallurgical Engineering from the Illinois Institute of Technology and the University of Toledo. Having worked in the heat treating and steel industries, he joined the heat treat equipment industry in 1966. During his 18 years at Midland Ross Corporation he was involved in research and design related to processing technologies for heat treatment—particularly carburizing, vacuum processing, and plasma technology, and holds patents related to these processing fields. Since joining Abar Ipsen Industries in 1985 as Technical Director he has been responsible for strategic planning of technology developments. In addition to this primary role, he oversees product development activities, maintains technology transfer with the German based Ipsen Industries, and serves as process consultant to both customers and the international group. He has also directed the development of a number of recent new products related to plasma carburizing, vacuum heat treating, and thermal cleaning.

ROSS E. PRITCHARD is a graduate Metallurgical Engineer from the University of Toronto and a registered Professional Engineer in the Province of Ontario, Canada. Involved in vacuum heat treating and brazing mainly in the aerospace and nuclear industries since 1955, Mr. Pritchard is Chairman and CEO of Vac-Aero International, Inc., Oakville, Ontario.

WILLIAM R. JONES, a Fellow of ASM International and member of several technical societies, holds numerous patents related to the field of instrumentation, electric power, and vacuum furnace technology. With a degree in Electrical Technology from Pennsylvania State University, he joined Manufacturers Engineering and Equipment Corporation of Warrington PA in 1955. There, he developed and applied electromechanical principles to industrial process control problems in trace moisture (dewpoint) and other analytical instrumentation, and designed an analog computer for close color matching of paints from instrumentation data. Later, at Instrument Development Laboratories, Attleboro MA, he applied the first two-color automatic optical pyrometer to crystal growing furnace, and investigated the application of automatic temperature control to many industrial furnace control problems. Moving to the Abar Corporation of Feasterville PA in 1962, he was involved in engineering design of electrical power supply, control circuit development, and hot zone design of high temperature vacuum furnace. Elected Vice President, Sales in 1964, Operations Vice President in 1967, and in 1973 became President of Abar Corporation. Mr. Jones is CEO of Vacuum Furnace Systems Corporation, a company he founded in 1978, and President of Solar Atmospheres, Inc., a commercial heat treating company he formed in 1985.

DAVID HOLKEBOER, a member of the American Vacuum Society and the American Society for Mass Spectroscopy, graduated from the University of Michigan with a Ph.D in Mechanical Engineering in 1961. As a vacuum systems engineer with the Aero Vac Corporation, he designed, built and installed custom high and ultrahigh vacuum systems, and also designed systems for the calibration of high vacuum gauges. Dr. Holkeboer is Manager of RGA Development for Lerybold-Inficon, Inc., and developed the first commercial microprocessor-based RGA, as well as simple, reliable sampling systems to extend the applicability of RGA instruments to partial vacuum and atmospheric pressure applications.

JAMES M. SULLIVAN, has a degree in Electrical Engineering from Northeastern University. He is a member of ASM International and is currently Chairman of the ASM Heat Treat Council. As his company's representative he serves as 2nd Vice President and a member of the Board of Directors of the Industrial Heating Equipment Association. He is also a member of the National Fire Protection Association. Mr. Sullivan began his 25-year career with Honeywell as a field engineer in Boston, and now is Manager, Industrial Heating Markets for the Industrial Automation and Controls Division of Honeywell.

WILBUR T. HOOVEN, III graduated from Duke University with a degree in Mechanical Engineering in 1960, is a member of ASM International and AWS, and over the years has written and presented a number of papers on practical vacuum process applications. Spending his early years as a sales engineer in the automotive and electronic equipment industries, in 1964 he became Division Manager for an international company in the metal processing job shop business. Mr. Hooven is President of Hooven Metal Treating, Inc., a company he founded in 1982. HMT is an aircraft quality commercial heat treater, specializing in vacuum brazing.

HOWARD I. SANDEROW, holds 2 patents, currently serves as Executive Director of the Center for P/M Technology, is a member of ASM International and other technical organizations, has written more than 25 technical papers in the P/M field, and is the author of a book on High Temperature Sintering. He has a BME from Rensselaer Polytechnic and an MBA from Wright State, and has 15 years of experience in the design and manufacture of powder metal components. Five years ago Mr. Sanderow formed Management and Engineering Technologies, an independent consulting firm specializing in the technology and business aspects of powder metallurgy and related particulate technologies such as ceramics and composites.

JEREMY ST. PIERRE received his diploma in Metallurgical Engineering from Ryerson Polytechnical Institute in Toronto, Canada and became a Development Metallurgist with the P/M division of Stackpole, Ltd., Canada.

In 1986 he joined C.I. Hayes, Inc. as a staff Metallurgist and has published numerous papers on high temperature sintering and vacuum carburizing. Mr. St. Pierre is Operations Manager with Hayes Heat Treating Corporation in Cranston RI.

JOHN H. DURANT, a 28 year member of ASM International, an instructor in vacuum metallurgy at New York University, cofounder and past Secretary of the American Vacuum Society, cofounder and past President of the Association of Vacuum Equipment Manufacturers, a member of various other technical societies, is currently a consultant in technology and industrial marketing development to the vacuum industry. The author of numerous technical papers, he has conducted seminars on vacuum thermal processing in the U.S., Eastern Europe, and China.

With a degree in Physical Sciences from Harvard College, Mr. Durant began his career in 1943 as a research assistant at the National Research Corporation involved in reduction process development; Equipment Division Engineer, vacuum distillation and dehydration; Business Manager, Research Division market development for continuous vacuum metallizing, vacuum melted bearing alloys, titanium, investment cast superalloys; and Vacuum Fusion Analyzer for residual gases in metals. As a Project Engineer at Engelhard Industries he was involved in nuclear fuel fabrication for research, naval and power reactors. Later, as Vice President of Marketing for Vacuum Industries, he was involved with products such as multiple purpose laboratory vacuum metallurgical processing system, vacuum sintering furnaces for dewaxing and sintering tungsten and mixed carbides, vacuum precision investment casting furnaces with directional solidization, vacuum annealing systems for titanium, zirconium and Inionel tubing, chemical vapor deposition equipment for ceramic coatings.

JEFFREY A. CONYBEAR, a member of ASM, graduated in 1990 from the University of Cincinnati with a degree in Metallurgical Engineering, and currently is pursuing a Masters degree in Engineering Management at Milwaukee School of Engineering. Mr. Conybear has worked in the heat treating industry, aluminum foundry, and die casting industry, and is Plant Metallurgist and Quality Engineer in charge of the ISO 9000 quality assurance program for Metal-Lab, Inc.

FOREWORD

In taking a hard look at vacuum technology, it became apparent to those associated with the industry that as more advances were made in vacuum processing a definite need existed for up-to-date information and practical answers to everyday problems; a useful reference for the furnace operator, the metallurgist, and the plant manager. This book is designed to fill that need. The information presented here has been contributed by industry experts whose collective knowledge and experience will be hard to find in any other place. It is the goal of the book to offer readers some effective problem-solving tools as well as an opportunity to widen their own experience.

Why vacuum? With the advent of space age materials and the need to process more and more sophisticated components, vacuum became the choice of heat treaters. As more was learned about the importance of the relationship of the surface of the component to the subsequent use of that component, it became increasingly more clear that vacuum processing was the only way to go. Because of all this, the vacuum processing side of the heat treatment industry has grown steadily. At the same time, vacuum has also opened new horizons for the joining and bonding of materials. Beyond titanium alloys, there is an ever-increasing number of highly-alloyed heat-resisting materials to be processed, with the space program and aircraft engine industry leading the way in vacuum processing.

What about vacuum equipment? Vacuum equipment is advancing rapidly and new materials and engineering advances have allowed the industry to break away from hot wall technology to embrace cold wall technology. Cooling methods now incorporate inert and high-pressure gas quenching as well as liquid quenching. Vacuum is now a partner on the shop floor.

What is the future of vacuum processing? The future of vacuum procressing will be its integration into the manufacturing cell. The vacuum furnace inline, processing parts within the time constraints that allow parts to be produced as needed. This means no inventory, and very short lead times. Will the industry be able to meet this challenge? I believe so. Furthermore, I am convinced that the vacuum processing industry must move in this direction if it is to become a competitive force within the world economy.

I would like to thank all the contributors whose knowledge and hard work have made this book a reality. A special thanks to my mentors over the years: George Bodeen, Robert Gilliland, Leo Thompson, and Lindberg Corporation who have made possible my active participation in ASM International.

Any reader who has questions or would like more information is invited to contact either myself or any of the authors.

Roger J. Fabian
Editor

Conversion Factors Useful in Vacuum Technology

Linear Measure

$\mathrm{mm} \times 0.3937 \times 10^{-1} = \mathrm{in.}$

$\mathrm{in.} \times 25.4 = \mathrm{mm}$

$\mu\mathrm{m} \times 39.370 = \mu\mathrm{in.}$

$\mu\mathrm{in.} \times 2.54 \times 10^{-2} = \mu\mathrm{m}$

$\mathrm{mil} \times 0.0254 = \mathrm{mm}$

$\mathrm{mm} \times 39.370 = \mathrm{mil}$

Volume

$\mathrm{m}^3 \times 10^3 = \mathrm{liter(l)}$

$\mathrm{l} \times 10^{-3} = \mathrm{m}^3$

$\mathrm{ft}^3 \times 2.832 \times 10^{-2} = \mathrm{m}^3$

$\mathrm{l} \times 0.353 \times 10^2 = \mathrm{ft}^3$

$\mathrm{in.}^3 \times 16.387 = \mathrm{cm}^3$

$\mathrm{cm}^3 \times 0.061 = \mathrm{in.}^3$

Weight/Mass

$\mathrm{lb} \times 4.536 \times 10^{-1} = \mathrm{kg}$

$\mathrm{kg} \times 0.220 \times 10^{-1} = \mathrm{lb}$

$\mathrm{oz} \times 2.835 \times 10^{-2} = \mathrm{kg}$

$\mathrm{kg} \times 3.527 \times 10^{-2} = \mathrm{oz}$

$\mathrm{tonne} \times 1.102 = \mathrm{ton}$

$\mathrm{ton} \times 0.907 = \mathrm{tonne}$

Stress/Force per Unit Area

$\mathrm{MN/m}^2 = \mathrm{MPa}$

$\mathrm{psi} \times 6.895 \times 10^{-3} = \mathrm{MPa}$

$\mathrm{ksi} \times 6.895 = \mathrm{MPa}$

$\mathrm{MPa} \times 0.145 = \mathrm{ksi}$

$\mathrm{kgf} \times 9.807 = \mathrm{MPa}$

$\mathrm{MPa} \times 0.102 = \mathrm{kgf}$

Pressure & Vacuum

$1\ \mathrm{atm} = 14.696\ \mathrm{psi} = 29.921\ \mathrm{in.Hg}$

$1\ \mathrm{atm} = 760\ \mathrm{mmHg} = 760\ \mathrm{torr}$

$1\ \mathrm{Pa} = 0.01\ \mathrm{millibar}$

$1\ \mathrm{millibar} = 0.76\ \mathrm{torr}$

$1\ \mathrm{bar} \times 10^5 = 1\ \mathrm{Pa}$

$1\ \mathrm{in.Hg}\ (32\ ^\circ\mathrm{F}) = 3.386 \times 10^3\ \mathrm{Pa}$

Temperature

$^\circ\mathrm{C} + 273.2 = {}^\circ\mathrm{K}$ (Kelvin)

$^\circ\mathrm{F} + 459.7 = {}^\circ\mathrm{R}$ (Rankine)

$(^\circ\mathrm{F} - 32) \times .55 = {}^\circ\mathrm{C}$ (Celsius)

$1.8\ ^\circ\mathrm{C} + 32 = {}^\circ\mathrm{F}$ (Fahrenheit)

$1.8\ ^\circ\mathrm{C} = {}^\circ\mathrm{F}$ (Interval)

$0.55\ ^\circ\mathrm{F} = {}^\circ\mathrm{C}$ (Interval)

Table of Contents

· 4 ·

· 5 ·

· 6 ·

· 10 ·

· 11 ·

• 1 •

Theory of Vacuum Technology

John R. Griffin and Sheldon W. Kennedy, Centorr/Vacuum Industries, Inc.

What is a Vacuum?

Most people accept the definition that any pressure lower than that of the atmosphere is a vacuum. The old international measurement system defines a standard atmosphere as the pressure required to raise a column of mercury to a height of 760 mm (29.9 in.), at a temperature of 0 °C (32 °F). The basic unit of the new international system of units (SI units) is the Pascal, or Newton per square meter, in which 101,325 Pascals equal 1 standard atm. Table 1.1 shows the pressure conversion factors for vacuum measurements on several scales. In the U.S. the most commonly used unit of pressure is the torr (named for Evangelista Torricelli, who invented the mercury barometer), which for practical purposes is the same as the millimeter (mmHg) and is the term most widely used in practical vacuum work. In Europe, the bar is more commonly used, but the scientific community is rapidly adopting

Table 1.1 Pressure conversion factors

To convert From / To	pascal (Pa)	torr or mmHg	millibar	atm	psi	in. Hg
	multiply by:					
pascal (N/m^2)	1	7.5×10^{-3}	10^{-5}	9.87×10^{-6}	1.45×10^{-4}	2.95×10^{-2}
torr (≈ mm of mercury)	133	1	1.33×10^{-3}	1.32×10^{-3}	1.93×10^{-2}	3.94×10^{-2}
millibar	100	0.75	1	9.87×10^{-4}	1.45×10^{-2}	29.5×10^{-3}
atm (normal atmosphere)	1.01×10^{5}	760	1.01	1	14.69	29.9
psi ($lb/in.^2$)	6.89×10^{3}	51.74	68.97×10^{-3}	6.80×10^{-2}	1	2.04
in. Hg (in. of mercury @ 0 °C)	3.39×10^{3}	25.4	3.39×10^{-2}	3.34×10^{-2}	0.491	1

the SI system. For convenience, and because this book is intended for practical use on the factory floor, the torr and the micron are used here.

Different pressure levels must be achieved by different means, which allow the different operations of interest to heat treaters to be performed:

1) Positive pressures from 2 to 6 bar (29.01 to 87.02 psia, or 14.31 to 72.32 psig): Pressure used for high-speed gas quenching. Positive pressures from 1 atm (14.696 psia) to 2 bar (29.01 psia) may be used in standard vacuum furnace chambers with special door clamping. The higher pressures above 2 bar require combination vacuum/pressure vessels designed and fabricated to the pressure vessel codes of the American Society of Mechanical Engineers (ASME).
2) Pressures from 1 bar to 6 bar absolute (750 torr to 4500 torr): Pressure to which inert gas is backfilled into the chamber at the completion of the process cycle for accelerated work cooling. Work-cooling blowers and heat exchangers operate in this range.
3) Pressures from 10^0 to 10^{-1} torr (1 torr to 100 microns): Pressures to which inert gas is backfilled when it is used to suppress vapors, commonly used for brazing and some sintering operations.
4) Pressures from 10^{-1} to 10^{-4} torr (100 microns to 0.1 micron): Pressure range in which graphite hot-zone furnaces operate at temperature when used for brazing, heat treating, and most hard metal and steel sintering.
5) Pressures from 10^{-3} to 10^{-5} torr (1 micron to .01 micron): Pressure range used for refractory metal hot-zone furnaces at temperature when used for diffusion bonding, heat treating, and brazing.
6) Pressures from 10^{-5} to 10^{-6} torr (0.01 micron to 0.001 micron). Pressure range at temperature for aluminum brazing, reactive metal sintering, and heat treating.
7) Pressures from 10^{-7} to 10^{-9} torr: Pressure range used for semi-conductor applications.

In general, the pressure range between 10^{-3} and 10^{-6} torr is referred to by the term "high vacuum."

Vacuum Pumping Systems

The vacuum pumping systems employed to attain the pressure levels desired for the applications listed above have one or more stages. These vacuum pumps are classified as roughing pumps and high-vacuum pumps. A roughing pump, also called a primary pump, is a mechanical rotary displacement pump in which a rotary piston or a rotor with sliding vanes compresses the air or gas at its inlet and discharges it to the atmosphere. The piston or vanes are sealed against the cylinder wall by a few drops per revolution of 30-wt hydrocarbon lubricating oil, recirculated from an internal reservoir. Figure 1.1 shows a cross section of a typical rotary piston pump. Single-stage mechanical pumps such as the one shown will maintain pumping speed approaching their displacement $ft^3/min.$ from atmosphere to about 1 torr pressure. Pumping speed declines rapidly, however, at lower pressures, and is effectively zero at 10^{-2} torr.

Most modern mechanical pumps are equipped with a gas ballast feature. When moist room air is compressed in the pump, its water vapor content condenses, which contaminates the oil in the pump. Water has a high vapor pressure compared with lubricating oil. Its presence adversely affects the ultimate pressure the pump will reach and, if allowed to build up in quantity, may cause damage to the pump by reducing the lubricating quality of the oil. One way to avoid water vapor condensation in the pump cylinder is to leak in enough dry air towards the end of the compression part of the cycle to reduce the resulting moisture level below that at which condensation can occur. This controlled leaking is called gas ballasting. It not only solves the condensation problem, but it also raises the ultimate pressure the pump can reach.

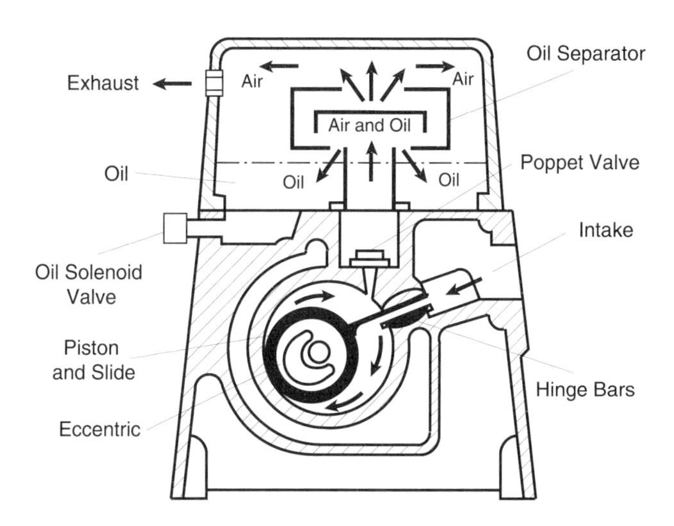

Fig 1.1 Rotary piston vacuum pump cross section

CAUTION: Vacuum furnaces sometimes produce or are used with hazardous or flammable gases. These gases concentrate in the pump to the danger level. Proper precautions are mandatory, including the treatment of gas ballast. See the pump manufacturer's manual for details.

Additional stages of gas compression are needed between the furnace chamber and the mechanical pump inlet, to improve the pumping speed for the broadest range of applications above. Added stages of compression may be provided either by a mechanical Roots blower, or by a diffusion pump, or both.

Figure 1.2 shows the use of a mechanical blower, which compresses a large volume of low-pressure gas to a smaller volume of higher-pressure gas for the inlet of the backing mechanical pump, which is more efficient at the higher pressures. Mechanical blowers reach maximum pumping speed in ft^3/min. in the 10^{-1} to 10^{-2} range, but their speed is zero at 3 to 5×10^{-3} torr.

To achieve pressures lower than 1×10^{-3} torr in the furnace chamber, a high-vacuum diffusion pump is used in addition to the blower or in place of it. This pump functions as illustrated in Fig 1.3. The silicone-based "oil" is raised to a temperature of approximately 275 °C (525 °F) by heaters mounted external to the base of the pump. The oil contained within the pump is vaporized and directed by the jets at a high velocity to the cooled pump wall, where it condenses and runs back to the boiler. Air molecules entering the

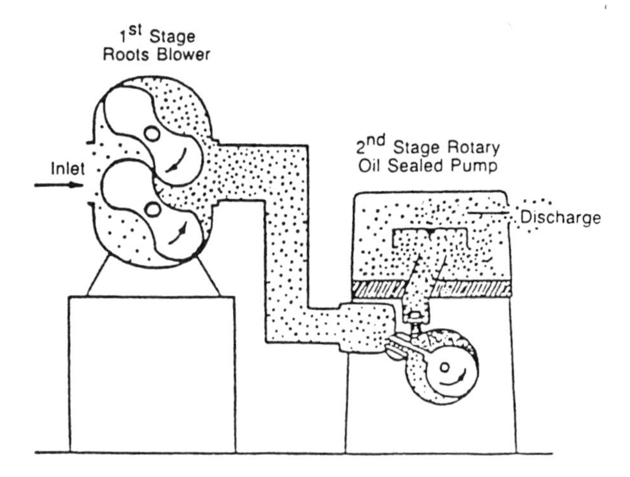

Fig 1.2 Mechanical blower and mechanical pump assembly

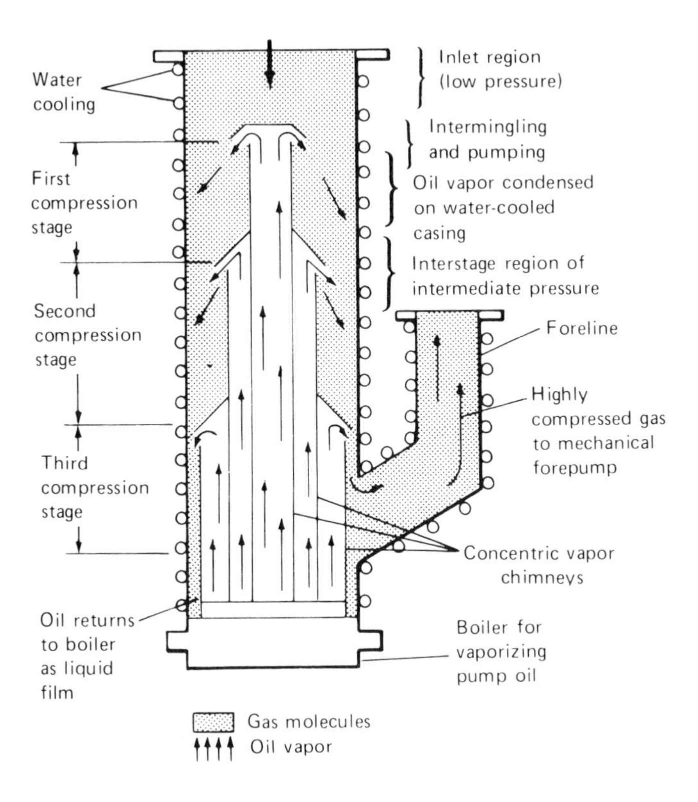

Fig 1.3 Oil diffusion pump

pump inlet are entrained by the vapor streams and are compressed in multiple stages to the pump outlet. The action is similar to that of a stream of ping pong balls being caught in a high-speed stream of golf balls.

The mechanical blower and/or mechanical pump must remove the air molecules from the diffusion pump outlet fast enough to achieve a pressure level no greater than 2×10^{-1} torr at the diffusion pump outlet, in order for the diffusion pump to work properly. At higher outlet (foreline) pressures, air molecules back-diffuse to the diffusion pump inlet as fast as they are pushed to the outlet.

To ensure proper foreline pressure, a small mechanical pump is opened to the diffusion pump foreline at all times. This "holding" pump assures proper foreline pressure even when the larger mechanical backing pump is being used for initial exhaust. It allows the diffusion pump to be turned on and operating even when the furnace is open.

CAUTION: Diffusion pumps are designed for operation with silicone-based fluids called "oils," which are fire- and explosion resistant. The use of other oils, especially hydrocarbons, is extremely hazardous. A failure in any part of the system which admits air or raises the foreline pressure will cause an explosion and fire.

In operation, exhaust of the vacuum vessel begins with the mechanical pump. When the pressure has been reduced enough for safe operation, the mechanical blower starts or a valve sequence opens the diffusion pump to the chamber while the mechanical pump backs the diffusion pump.

Furnace manufacturers design pumping systems to be the most effective and least expensive in their judgment. In designing a vacuum pumping system, the basic consideration is that the through-put is the same in all parts of the system at any given instant. Through-put is determined by pumping speed times pressure. To illustrate, a system may require a 16-inch diffusion pump. Its maximum through-put occurs at 1×10^{-3} torr (1 micron) pressure, and its rated speed is approximately 5000 liters/s (10,593 ft^3/min.) when equipped with an inlet baffle. Maximum through-put is therefore 10,593 micron-ft^3/min. The backing pump must also have a through-put capability of 10,593 micron-ft^3/min., but the diffusion pump will compress the gas to 200 microns (2×10^{-1} torr) at its foreline. The minimum speed required for the backing pump is therefore 10,593/200 = 53 ft^3/min. at an inlet pressure of 2×10^{-1} torr. Few single-stage mechanical pumps, operating with gas ballast, can achieve this speed at 2×10^{-1} torr inlet pressure. For this reason, mechanical blowers are frequently used to back the diffusion pump and are, in turn, backed by the mechanical pump. The smallest standard mechanical blower has a rated inlet speed of 100 ft^3/min. at 2×10^{-1} torr pressure, resulting in a through-put capability of 20,000 micron-ft^3/min., which is well above the minimum required for the diffusion pump. It may be backed by a small, single-stage mechanical pump operating with gas ballast. A pump of 30 ft^3/min. displacement will meet the 10,593 micron-ft^3/min. through-put criterion at an inlet pressure of 3 torr. The blower output pressure will be 3 torr when the inlet pressure is 2×10^{-1} torr. The small blower and pump described are more than adequate to back the diffusion pump at its maximum through-put.

A second consideration in the design of a pumping system is the time required to pump out the chamber to enable heating to be started. The 30 ft^3/min. pump and blower discussed above require about 22 min. to exhaust a 5 ft diameter × 5 ft long vessel (about 100 ft^3) to the pressure of 2×10^{-1} torr, at which the diffusion pump may be opened. This time is reduced to 5 min.

by the use of a blower of 245 ft^3/min. speed backed by a pump of 150 ft^3/min. displacement. Use of these simple calculations, along with the pump speed data supplied by the manufacturers, allows the most appropriate pumping system for any furnace to be designed or revisions to be made to an older furnace.

Measuring Vacuum Levels

Pressures in vacuum vessels such as furnace chambers may be measured from above atmospheric pressure to a few torr by Bourdon-tube or capsule-dial gages. At lower pressures, however, such gages do not have sufficient resolution, and therefore electrical gages, which infer the pressure indirectly, are commonly used for these lower pressure ranges. They measure the total gas and vapor and also provide an electrical signal which can be recorded or used for control.

The most inexpensive and common gage for the 20 torr to 1×10^{-3} torr range is the thermocouple vacuum gage. The inside of a small metal canister containing the sensing elements is exposed to the vacuum to be measured. The canister contains a wire through which a constant current flows from the gage control. The wire becomes cooler or hotter depending on the heat conducted away from it by the gas or vapor in the system. The more gas, the cooler the wire. Wire temperature is sensed by a thermocouple and its output is displayed on a meter scale calibrated in pressure units. The operating range of these gages is limited because at pressures below 10^{-3} torr the thermal conductivity of the gases is too small to measure by this means, and at pressures above 20 torr the thermal conductivity of the gas is nearly constant, preventing accurate readings.

Capacitance manometers are more expensive but are coming into wider use as vacuum gages because they can cover wider ranges than other types and produce an electrical output. They consist of an electrical capacitor, one side of which is a very thin metal membrane exposed on one side to the vacuum to be measured. Deflection of the diaphragm caused by pressure changes is reflected as a change in electrical capacitance. This variable capacitor is part of a capacitance bridge circuit which amplifies the signal and supplies an electrical output that can be displayed or recorded. This type of vacuum gage can be made to cover the pressure range from atmosphere to 10^{-6} torr in a few stages, and it is not affected by the thermal conductivity of the gas.

Pressures of 1×10^{-3} torr and lower are usually measured by ionization gages, in which the gage elements are enclosed in a glass bottle or metal canister and exposed to the vacuum to be measured. The control provides power to either a hot filament or a cold cathode which emits electrons that collide with gas molecules, knocking off other electrons and leaving positive ions. The positive ions are collected on a negatively charged plate, and the ion current becomes an absolute and direct measure of the number of gas molecules in the volume, and hence, of the pressure.

With the gages described above, the entire pressure range from atmosphere down to the lowest achievable can be measured, recorded, and used for control.

Heat Transfer in Resistance Furnaces

Heat is generated by electric resistance elements in most vacuum furnaces used for heat treating. While some furnaces use induction heaters, they are initially more expensive, and induction heat is difficult to control for uniform heating results.

In the absence of air or a gas atmosphere, conduction heating and convection currents cannot occur. Heat transfers from the hot resistance elements to the cooler workload only by radiation. It transfers from the hot body to the cold in accordance with the 4th power of their relative absolute temperatures. In the most simplified form: Heat transfer per unit time = ($T_1 4$ - $T_2 4$), where T_1 and T_2 are absolute temperatures in either degrees Kelvin (°C + 273) or degrees Rankine (°F + 459). From this mathematical expression

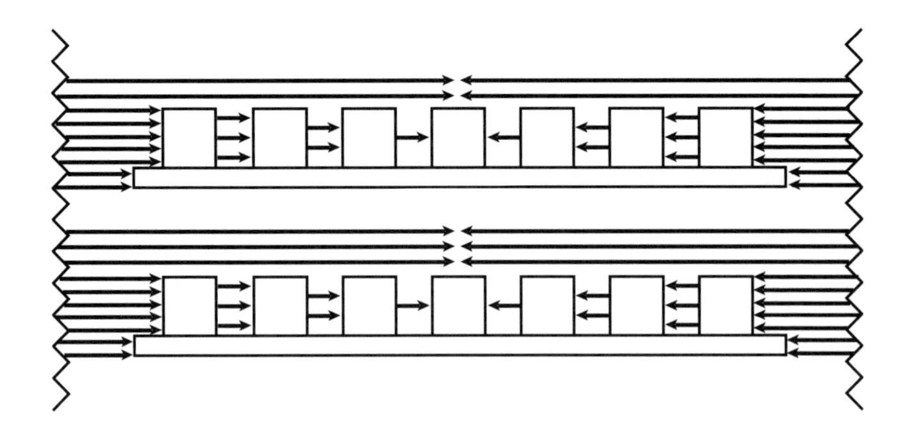

Fig 1.4 Heating multiple workload pieces by radiation. Note shadowing of pieces in the center of the load

it can be seen that heat is transferred more efficiently at large differences between T_1 and T_2, as well as at higher temperatures.

The element temperature, T_1, is lowered by transfer of heat to the work. The furnace temperature sensors detect this change and signal the temperature controller to add heat to the elements until the work temperature, T_2, reaches the desired level.

Heat radiation is a straight-line phenomenon the same as light, per Fig 1.4, and the resistance elements radiate in all directions, as shown in Fig 1.5. When a furnace contains a workload comprising many individual workpieces, the outer rows of them nearest the elements are heated by direct radiation from the heating elements, plus reflection from the hot face of the furnace insulation. The outer row, in turn, radiates to the next row, and so on to the center of the workload. The center of a dense workload therefore arrives at the desired temperature much more slowly than the outer rows. The actual rate of heating of the center of a densely packed workload is difficult to calculate, involving as it does the emissivity of the workpieces, the heat conductivity of the work and the shelves, the space for radiant heat between the workpieces and the shelves, etc. Good practice is discussed in subsequent chapters, but load thermocouples are recommended to record the exact heating rate for a given load.

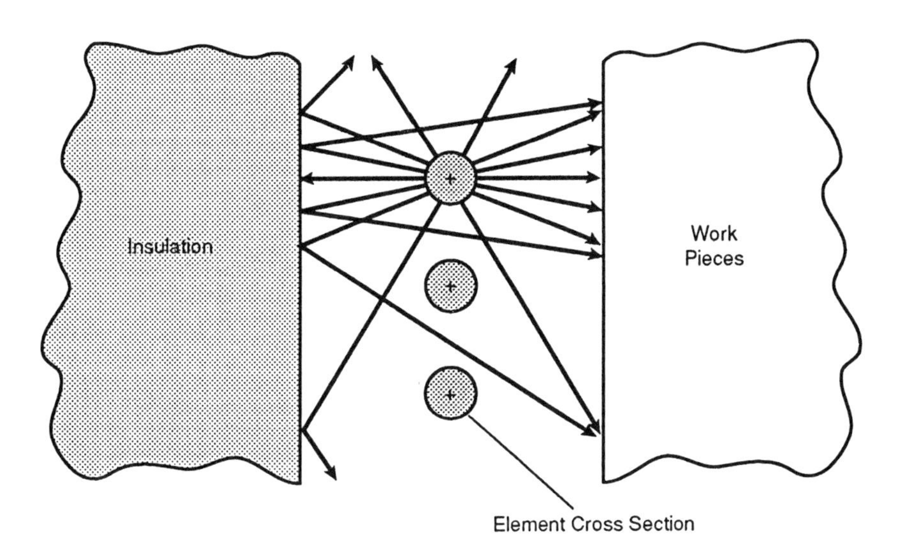

Fig 1.5 Radiation pattern of rod elements

When the resistance elements are rods, tubes or slats, a plane that defines the outer limits of the uniform temperature zone is located a short distance from the plane of the elements, per Fig 1.6. Also, in a furnace with elements surrounding the work on four sides but not on the ends, heating of the end face of the work depends on radiation from the last elements on the sides, top, and bottom. The plane that defines the uniform heat zone at the ends is shown in Fig 1.7.

The uniform temperature zone, or effective hot zone, of the furnace can be thus defined, bounded by imaginary lines in space. Except for small laboratory furnaces, furnace makers specify furnace sizes by their effective hot zone dimensions, and the resistance elements are set back as shown. The workload should be confined to the uniform zone and loading close to either the elements or to the unheated insulation pack should be avoided. Note the cool zone near the end shields. Some heat treaters call this the "black zone."

The location of the temperature sensor is a critical issue in furnace use. The sensor, usually a sheathed thermocouple, is best located near the work but in the path of direct radiation from the elements. The sensor then senses an average between the hot elements and the cool work. Its signal is used by the controller as part of the instruction to the furnace power supply. The

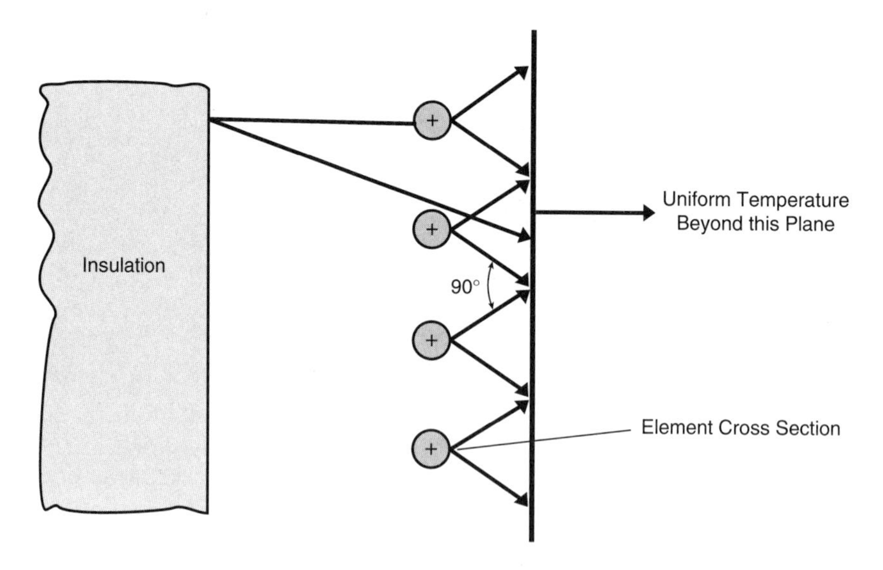

Fig 1.6 Distance of uniform temperature plane from rod element array

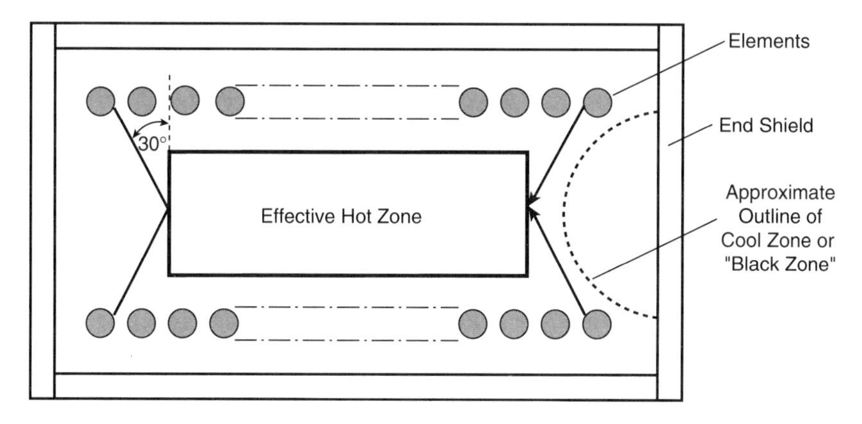

Fig 1.7 Minimum extension of elements beyond effective hot zone for uniform temperature across end face of zone

controller seldom calls for full power output because at any given instant it is comparing the heating ramp temperature command (entered into the programmer), with the thermocouple output, which is influenced by the element temperature. On the other hand, if the thermocouple is installed out of sight of the elements and, for example, within the workload, it sees only the work temperature and calls for more and more heat because heat penetration to the work center is slow. The result may be overheating or even melting of workpieces in the outer rows.

Work Cooling and Gas Quenching

Cooling the workload in a furnace under vacuum is a slow process. The heat in the load must pass by radiation through the heat insulation to the cold chamber wall. Cool-down times under vacuum are totally dependent upon the thermal efficiency of the insulating package surrounding the work, and cooling usually requires many hours.

Backfilling the furnace to 650 to 700 torr pressure with a gas that is inert to the work, such as argon or nitrogen in most cases, speeds cooling even if the gas is still and not blown about. Many systems are designed to accommodate several atmospheres of gas when higher cooling rates are required. The gas, which conducts the heat to a heat exchanger or to the cold wall of the chamber, shortens cooling times significantly. The heat loss through the furnace insulation when argon or nitrogen are backfilled is about 150% of the loss in vacuum, 175% for helium, and 200% for hydrogen.

Faster cooling is accomplished by recirculating the gas between the work and a heat exchanger at a speed of a few hundred ft^3/min. A motorized blower installed inside the furnace chamber but outside the furnace hot zone performs this forced convection. Shutters or bungs are opened in the hot zone at the time of cooling to allow an unobstructed flow of cooling gas over the work and the heat exchanger surface. Figure 1.8 illustrates relative cooling times. The fastest cooling short of immersion in a liquid bath is accomplished by a gas quench system which is an integral part of the furnace.

A basic equation governing gas quenching is (Ref 1):

$$t_1 = \frac{W\,C_p}{A\,h_1}\, l_n\, \frac{T_1 - T_f}{T_2 - T_f}$$

where, t = Time in hours from T_1 to T_2

W = Weight of load in pounds
C_p = Specific heat of load — btu/lb — °F
A = Surface area of workload — square feet
h_1 = Heat transfer coefficient — load to fluid (gas) — btu/hr — ft^2 — °F
T_f = Average fluid (gas) temperature — °F
T_1 = Initial load temperature — °F
T_2 = Final load temperature — °F

From the equation it is evident that cooling time is shortened by reducing the weight of the work or by increasing its surface area, but these factors are seldom negotiable, since they are defined by the specific workload. Increasing factor h1 will also shorten cooling time. It can be raised significantly by the gas quench system design. Increasing the mass velocity of the gas (expressed in $lb/h\text{-}ft^2$) over the work and the heat exchanger increases factor h_1 by a large factor. The mass flow is increased both by increasing the linear velocity of the gas and increasing its mass, by increasing its pressure. A turbine-type blower provides maximum volume and velocity. The pressure may be raised to as much as 2 bar (29.01 psia) in standard vacuum furnaces with clamped doors, to increase the mass flow by a factor of 1.5 to 1.7. If a special ASME code-qualified vessel is employed, the pressure may be increased to as much as 6 bar (72.52 psia), with an increase in mass flow by a factor of about 4.0.

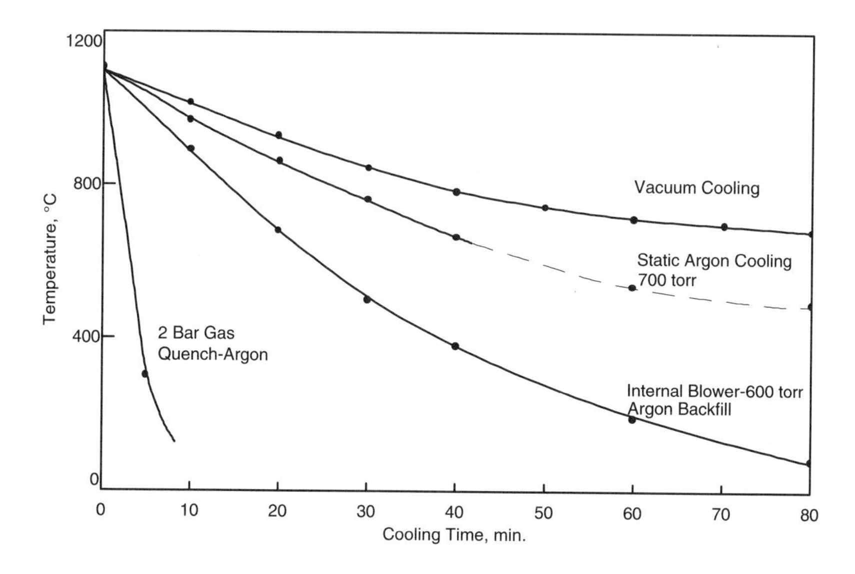

Fig 1.8 Typical cooling curves for intermediate size vacuum furnaces with metal hot zones and normal workloads

A cooling gas jet directed perpendicular to the work face has been shown to be much more efficient than parallel flow along the surface. Cooling gas jets are arranged, therefore, to direct the gas against the work. They are kept as close to the work as practical, to avoid diffusion in a large volume before impingement on the work. Figure 1.9 illustrates the arrangement in a typical heat treating furnace. While jets of cooling gas which impinge on the work face have been shown to be more efficient than a parallel flow of gas along the work surfaces, impinging cooling jets may distort thin workpieces.

Of course, the heat from the work must be transferred to the cooling water through a gas-water heat exchanger. The equation also shows that lowering the average gas temperature by the use of a large-capacity heat exchanger will improve the cooling time. Heat exchanger capacities ranging from 250,000 to several million Btu/h are used, depending on the size of the furnace and the load and quench speed required.

Use of very cold cooling water lowers the average fluid temperature, T_f, and improves the gas quench rate. Very cold water in the heat exchanger and furnace chamber, however, causes moisture condensation (sweating) during

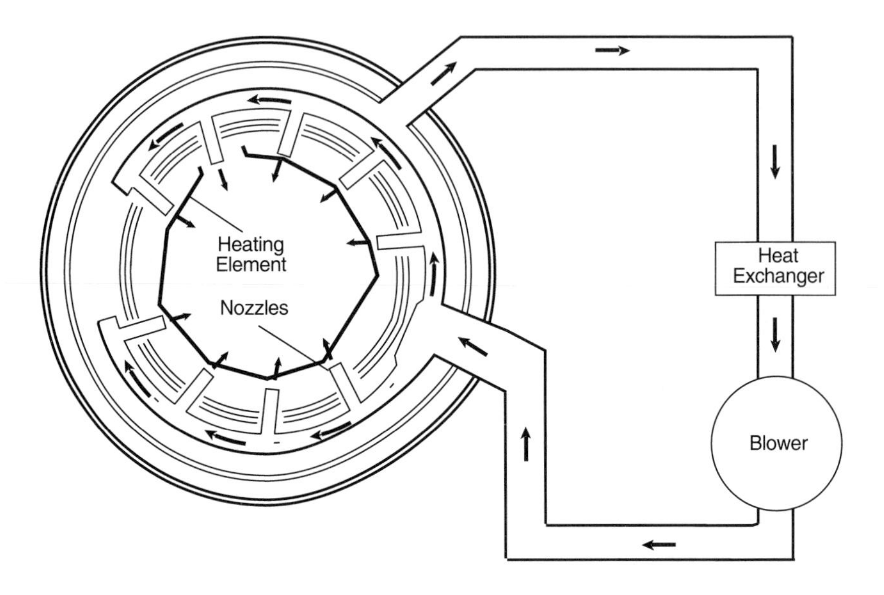

Fig 1.9 Gas quench system with impinging flow

humid summer weather, with consequent poor vacuum pumping performance. Heat exchangers designed for a very large flow of warmer water are a more practical solution.

As is seen from the foregoing, designing a gas quench system for any specific application is a complex procedure. Many case studies exist to demonstrate that gas quenching can be performed in seconds at speeds closely approximating the speed of an oil quench.

Vapor Pressure

Vapor pressure, defined as the gas or vapor pressure exerted when a substance is in equilibrium with its own vapor, depends solely on the temperature of the substance. For instance, imagine a liquid placed in an exhausted container and maintained at a specified temperature. Evaporation takes place and just enough liquid evaporates to build up the pressure of the vapor corresponding to the temperature at which the system is maintained.

In the case of solid substances, some molecules at the surface have higher energy than others and escape as vapor. If the substance is in an evacuated enclosure where there are no other gases or vapors, molecules will continue to escape until their rate of escape is balanced by their rate of recapture or return to the surface. The equilibrium pressure developed is the vapor pressure of that substance at that temperature. If the vapor molecules are pumped away but the temperature maintained, more molecules will escape from the solid surface in an attempt to build the pressure back up to equilibrium at that temperature. Thus, evaporation proceeds rapidly at pressures lower than equilibrium. Conversely, if the pressure in the enclosure is raised by the introduction of other gas molecules, the rate of evaporation is slowed.

Table 1.2 shows vapor pressures of the elements. Several metallic oxides are also found in vacuum furnaces, and Table 1.3 shows vapor pressures of a few common oxides.

Vapor pressure considerations are most significant in vacuum furnace operation. For instance, zinc, a common alloy of brass, vaporizes at such low temperatures that brass probably should be avoided for use within a vacuum furnace. Manganese and chromium also evaporate at relatively low temperatures. To avoid depletion of these materials at heat treating temperatures, the pressure level surrounding the workpieces must be maintained above the diffusion pumping level. A low-pressure backfill or partial pressure control system of inert gas is usually provided for this purpose.

The vapor pressure of most brazing alloys is also relatively high at the melting point of the alloy. Depletion of the alloy can be avoided by raising the pressure in the furnace to a level above the vapor pressure of the alloy at brazing temperature. Braze alloy vapor pressure data is usually available from the alloy supplier. If not, but if the composition of the alloy is known, Ref 2 suggests that the vapor pressure of each constituent can be estimated from the equation:

$$1_n P = 10.56 \frac{T - T_b}{T}$$

where, T = Temperature of interest — K

T_b = Boiling point of the material — K

P = is in Torr

The total vapor pressure of the braze alloy is the sum of the vapor pressures of each constituent times its percent in the alloy. Figure 1.10 shows the vapor pressures of several common brazing alloy constituents.

Table 1.2 Density and vapor pressure of common elements

Element	Symb.	At. No.	Dens. (S.G.)	At. Wt.	Melt. Pt.	Temp. °C at which Vapor Pressures are: .001mm	.01mm	0.1mm	1.0mm	10mm	100mm	760mm
Aluminum	Al	13	2.7	27.0	660	889	996	1123	1279	1487	1749	2327
Antimony	Sb	51	6.7	121.8	630	595	678	779	904	1033	1223	1617
Arsenic	As	33	5.7	74.9	817	237	277	317	362	437	517	613
Barium	Ba	56	3.5	137.4	717	625	721	840	961	1049	1301	1613
Beryllium	Be	4	1.9	9.0	1284	1029	1212	1367	1567	1787	2097	2507
Bismuth	Bi	83	9.8	209.0	271	609	698	802	934	1136	1271	1627
Boron	B	5	2.3	10.8	2300	1239	1355	1489	1648	3030	3460	2527
Cadmium	Cd	48	8.6	112.4	321	220	264	321	394	484	611	765
Calcium	Ca	20	1.5	40.1	810	528	605	700	817	983	1207	1482
Carbon	C	6	2.0	12.0	3700	2471	2681	2926	3214	3946	4373	4552
Cerium	Ce	58	6.9	140.1	640	1190	1305	1439	1599	1708	2018	2527
Cesium	Cs	55	1.9	132.9	29	110	153	207	277	387	515	690
Chromium	Cr	24	6.9	52.0	1900	1090	1205	1342	1504			2222
Cobalt	Co	27	8.9	58.9	1478	1494	1649	1833	2056	2380	2720	3097
Copper	Cu	29	8.9	63.5	1083	1141	1273	1432	1628	1879	2207	2595
Gallium	Ga	31	5.9	69.7	30	965	1093	1248	1443	1541	1784	2427
Germanium	Ge	32	5.3	72.6	959	1112	1251	1421	1635	1880	2210	2707
Gold	Au	79	19.3	197.0	1063	1316	1465	1646	1867	2154	2521	2966
Indium	In	49	7.4	114.8	157	840	952	1088	1260	1466	1756	2167
Iodine	I	53	4.9	126.9	114	–31	–11	12	39	72	115	183
Iridium	Ir	77	22.4	192.2	2454	2340	2556	2811	3118			4527
Iron	Fe	26	7.9	55.9	1535	1310	1447	1602	1783	2039	2360	2727
Lanthanum	La	57	6.2	138.9	826	1242	1381	1549	1754	1998	2330	2727
Lead	Pb	82	11.3	207.2	328	625	718	832	975	1167	1417	1737
Lithium	Li	3	0.5	6.9	179	439	514	607	725	890	1084	1367
Magnesium	Mg	12	1.7	24.3	651	383	443	15	605	702	909	1126
Manganese	Mn	25	7.2	54.9	1244	878	980	1103	1251	1505	1792	2097
Mercury	Hg	80	13.5	200.6	–40	18	48	82	126	184	216	361
Molybdenum	Mo	42	10.2	96.0	2622	2295	2533	2880	3102	3535	4109	4804

(continued)

Table 1.2 Density and vapor pressure of common elements

Element	Symb.	At. No.	Dens. (S.G.)	At. Wt.	Melt. Pt.	Temp. °C at which Vapor Pressures are: .001mm	.01mm	0.1mm	1.0mm	10mm	100mm	760mm
Neodymium	Nd	60	6.9	144.3	840	1192	1342	1537	1775	2095	2530	3090
Nickel	Ni	28	8.9	58.7	1455	1371	1510	1679	1884	2007	2364	2837
Osmium	Os	76	22.5	190.2	2697	2451	2667	2920	3221	3946	4380	4627
Palladium	Pd	46	12.0	106.4	1555	1405	1566	1759	2000	2280	2780	3167
Phosphorus	P	15	1.8	31.0	597	160	190	225	265	310	370	431
Platinum	Pt	78	21.5	195.	1774	1904	2090	2313	2582	3146	3714	3827
Potassium	K	19	.9	39.1	64	161	207	265	338	443	581	779
Rhenium	Re	75	20.5	186.2	3180	2790	3060	3400	3810			5630
Rhodium	Rh	45	12.1	102.9	1967	1971	2149	2358	2607	2880	3392	3877
Rubidium	Rb	37	1.5	85.5	39	123	165	217	283	387	519	679
Ruthenium	Ru	44	8.6	101.1	2427	2230	2431	2666	2946			4227
Scandium	Sc	21	2.5	45.0	1200	1282	1423	1595	1804			2727
Selenium	Se	34	4.3	79.0	217	200	235	280	350	430	550	685
Silicon	Si	14	2.4	28.1	1410	1223	1343	1585	1670	1888	2083	2477
Silver	Ag	47	10.5	107.9	961	936	1047	1184	1353	1575	1865	2212
Sodium	Na	11	0.9	23.0	98	238	291	356	437	548	696	914
Strontium	Sr	38	2.6	87.6	771	475	549	639	750	898	1111	1384
Sulfur	S	16	2.1	32.1	119	66	97	135	183	246	333	444
Tantalum	Ta	73	16.6	180.9	2996	2820	3074	3370	3740			6027
Tellurium	Te	52	6.2	127.6	450	336	383	438	520	633	792	990
Thallium	Tl	81	11.9	204.4	304	527	606	702	821	983	1196	1457
Thorium	Th	90	11.2	232.0	1827	1999	2196	2431	2715			4227
Tin	Sn	50	5.7	118.7	232	1042	1189	1373	1609	1703	1968	2727
Titanium	Ti	22	4.5	47.9	1820	1384	1546	1742	1965	2180	2480	3127
Uranium	U	92	18.7	238.1	1132	1730	1898	2098	2338			3527
Vanadium	V	23	5.9	51.0	1697	1725	1888	2079	2207	2570	2950	3527
Yttrium	Y	39	5.5	88.9	1477	1494	1649	1833	2056	2202	2539	3227
Zinc	Zn	30	7.1	65.4	419	292	343	405	487	593	736	907
Zirconium	Zr	40	6.4	91.2	2127	1818	2001	2212	2459			3577

Table 1.3 Vapor pressures of various oxides

Compound	Formula	Temperature °C (°F) at which Vapor Pressures are: 10^{-5}mm	10^{-4}mm	10^{-3}mm	10^{-2}mm	10^{-1}mm	760mm
Aluminum oxide	Al_2O_3	...	1880 (3390)	1980 (3570)	2105 (3790)	2265 (4080)	3545 (6330)
Chromic oxide	Cr_2O_3	1320 (2410)	1425 (2600)	1555 (2830)	1695 (3080)	1875 (3380)	3030 (5440)
Magnesium oxide	MgO	1390 (2530)	1510 (2750)	1630 (2970)	1790 (3250)	1955 (3550)	3630 (6510)
Molybdenum trioxide	MoO_3	...	...	595 (1100)	625 (1160)	700 (1290)	795 (1460)
Silicon dioxide	SiO_2	1365 (2490)	1480 (2700)	1605 (2920)	1750 (3160)	1910 (3440)	2235 (4050)
Titanium oxide	TiO_2	...	1575 (2870)	1705 (3100)	...	...	...
Iron oxide	Fe_2O_3	...	...	...	1025 (1880)	...	...

Note: The vapor pressure of Fe_2O_3 at 435 °C (1700 °F) is 4×10^{-4} mm and the vapor pressure of FeO at 1135 °C (2060 °F) is 7×10^{-11}. Iron oxide is complex with respect to vapor pressure and reactions are dependent upon the formula and the ability to maintain nonequilibrium conditions in favor of vacuum to sweep away the oxide molecules.

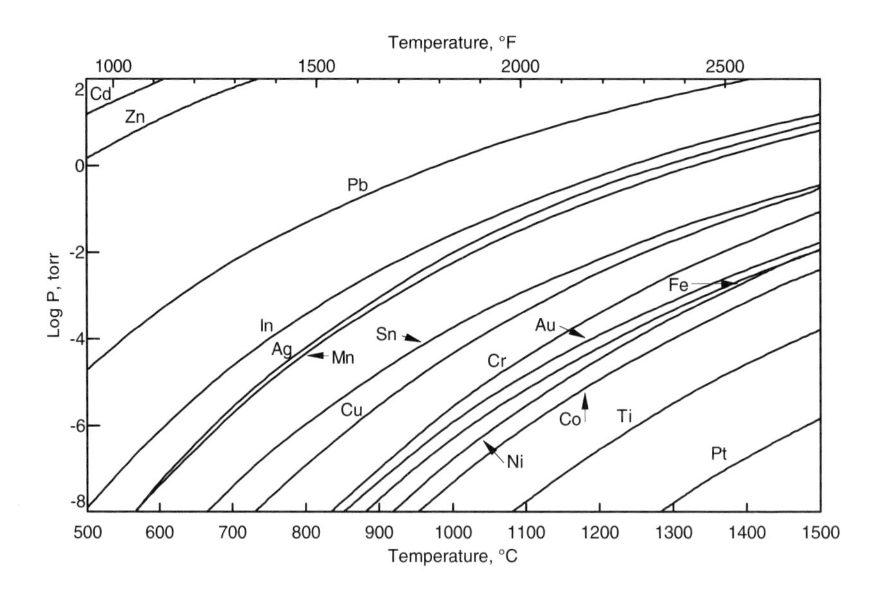

Fig 1.10 Vapor pressure of braze alloy constituents

Oxidation and Reduction

One of the primary reasons for the use of vacuum furnaces is to limit oxidation. Results from vacuum furnaces are consistent so long as pressure and temperature cycles are consistent and leaks are avoided. Vacuum pumping reduces oxygen level at the workpieces, even if the source of oxygen is water vapor adsorbed onto the part surface and furnace structure. Table 1.4 shows the dew point equivalents of several vacuum levels.

Any oxygen remaining after vacuum pumping begins will react spontaneously with the work at relatively low temperatures. For instance, chromium, which imparts the rust-free characteristic to stainless steel, begins to oxidize at about 370 °C (700 °F). Molybdenum begins to oxidize at about 325 °C (620 °F) and carbon (graphite) at about 260 °C (500 °F). The kinetics of the reaction are usually slow at the low temperatures, but they increase rapidly with temperature.

Once formed, many oxides are difficult to dissociate in a vacuum furnace. Figure 1.11 shows the relative stability of several oxides, which illustrates the relative instability of copper oxide but the great stability of alumina. The pressure at which the oxides dissociate can be calculated from the free energy of formation shown in the figure. For instance, what is the oxygen pressure in equilibrium with copper and copper oxide at 980 °C (1800 °F)? From the diagram:

$$G^\circ \text{ (cal)} = -37 \text{ kcal/mol } O_2 \text{ at } 1255 \text{ K}$$

$$G^\circ = -RT\ 1_n\ K = RT\ 1_n P_{O_2} = -37{,}000 \text{ cal.}$$

$$\frac{-37{,}000}{1.987 \times 1255}$$

$$P_{O_2} = e = 3.6 \times 10^{-7} \text{ atm} = 2.7 \times 10^{-4} \text{ Torr}$$

where,

T = Absolute temperature - K

R = Gas constant - 1.987 cal/mol - K

While most vacuum furnaces will reach a pressure low enough to allow copper oxide to dissociate, the same calculation shows that stable chromium oxide, Cr_2O_3, will require a pressure of 3×10^{-13} torr to dissociate at 1315 °C (2400 °F)! As a practical matter, however, Cr_2O_3 will evaporate rapidly at this temperature and a pressure of 1×10^{-5} torr. Nonetheless, these figures explain the reason that heavily oxidized stainless steel fixtures can be used for copper brazing. The copper does not stick to the chromium oxide, which remains stable at copper brazing temperatures of about 1135 °C (2075 °F), while the copper oxide has disappeared.

Table 1.4 Dewpoint equivalent to water vapor pressure

Dewpoint °C	Dewpoint °F	Vapor Pressure (Water/Ice in Equilibrium) mm of Mercury	Dewpoint °C	Dewpoint °F	Vapor Pressure (Water/Ice in Equilibrium) mm of Mercury
–150	–238	7×10^{-15}	–42	–44	.0768
–140	–220	3×10^{-10}	–40	–40	.0966
–130	–202	7×10^{-8}	–38	–36	.1209
–120	–184	10×10^{-8}	–36	–33	.1507
–118	–180	.00000016	–34	–29	.1873
–116	–177	.00000026	–32	–26	.2318
–114	–173	.00000043	–30	–22	.2859
–112	–170	.00000069	–28	–18	.351
–110	–166	.0000010	–26	–15	.430
–108	–162	.0000018	–24	–11	.526
–106	–159	.0000028	–22	–8	.640
–104	–155	.0000043	–20	–4	.776
–102	–152	.0000065	–18	0	.939
–100	–148	.0000099	–16	+3	1.132
–98	–144	.000015	–14	+7	1.361
–96	–141	.000022	–12	+10	1.632
–94	–137	.000033	–10	+14	1.950
–92	–134	.000048	–8	+18	2.326
–90	–130	.000070	–6	+21	2.765
–88	–126	.00010	–4	+25	3.280
–86	–123	.00014	–2	+28	3.880
–84	–119	.00020	0	+32	4.579
–82	–116	.00029	+2	+36	5.294
–80	–112	.00040	+4	+39	6.101
–78	–108	.00056	+6	+43	7.013
–76	–105	.00077	+8	+46	8.045
–74	–101	.00105	+10	+50	9.209

(continued)

Table 1.4 Dewpoint equivalent to water vapor pressure

Dewpoint °C	Dewpoint °F	Vapor Pressure (Water/Ice in Equilibrium) mm of Mercury	Dewpoint °C	Dewpoint °F	Vapor Pressure (Water/Ice in Equilibrium) mm of Mercury
–72	–98	.00143	+12	+54	10.52
–70	–94	.00194	+14	+57	11.99
–68	–90	.00261	+16	+61	13.63
–66	–87	.00349	+18	+ 64	15.48
–64	–83	.00464	+20	+ 68	17.54
–62	–80	.00614	+22	+ 72	19.83
–60	–76	.00808	+24	+ 75	22.38
–58	–72	.0106	+26	+ 79	25.21
–56	–69	.0138	+28	+ 82	28.35
–54	–65	.0178	+30	+ 86	31.82
–52	–62	.0230	+32	+ 90	35.66
–50	–58	.0295	+34	+ 93	39.90
–48	–54	.0378	+36	+ 97	44.56
–46	–51	.0481	+38	+100	49.69
–44	–47	.0609	+40	+104	55.32
			+42	+108	61.50
			+44	+111	68.26

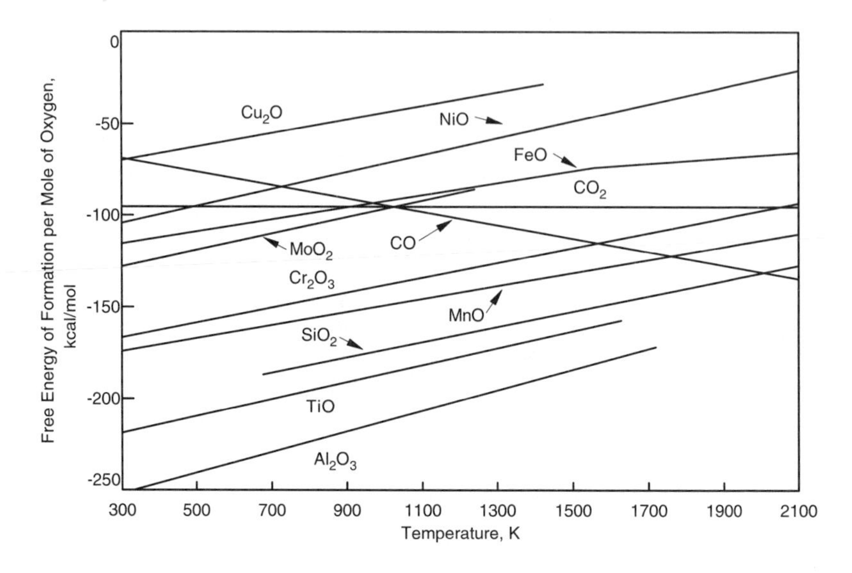

Fig 1.11 Oxide stability diagram

Note also the stability of carbon monoxide on the diagram. At elevated temperatures it is more stable than many metallic oxides. When graphite is present in a vacuum furnace, the carbon unites with any oxygen from the remaining air or water vapor to form CO, which remains stable and is pumped away. In this way, it lowers the oxygen content of the remaining gas to minimize oxidation of the workpieces.

Conclusion

The foregoing touches on a part of the technology employed when vacuum heat-treating furnaces are made or used. Succeeding chapters enlarge on several aspects of the technology as it affects the user. In addition, the reference works and the works listed in the bibliography provide a depth of detail on the scientific basis for the technology.

One final word of caution with respect to safety: Although a vacuum furnace appears to be harmless so long as the electrical and mechanical hazards in operation are avoided by carefully following the manufacturer's instructions, the use of gases is hazardous. Most applications, at some point in the floor-to-floor cycle, either use or generate colorless, odorless gases which, in sufficient concentrations, will cause asphyxia when breathed. Asphyxia causes immediate unconsciousness, and death may follow in minutes. Operators and maintenance personnel must not enter, lean over into, or climb up into furnaces that have been backfilled with argon or nitrogen until the oxygen adequacy of the environment in the furnace has been proved. Small, safe, concentrations of hydrogen are sometimes used in furnaces. Hydrogen will, however, concentrate to an explosive level in mechanical pumps. Carbon monoxide, generated by some processes, is poisonous in concentrations above 50 ppm in air. In both cases proper venting of the pumps and proper use of the gas ballast feature is mandatory to avoid accidents.

References

1) K.W. Doak, Custom design using helium..., *Heat Treating*, Vol. XVII, No. 10, Sept 1985

2) E.T. Thompson and C.W. Finn, Vacuum Brazing Technology, *Industrial Heating*, June 1988

Selected References

1) C.M. Van Atta, *Vacuum Science and Engineering*, McGraw-Hill, 1965

2) W.H. McAdams, *Heat Transmission*, McGraw-Hill, 1954

3) Kubaschewski, Evans, and Alcock, *Metallurgical Thermochemistry*, Pergamon Press, London, 1967

• 2 •

Introduction to Vacuum Thermal Processing

Carmen Paponetti, HI TecMetal Group

A relatively new development in the metallurgical field, vacuum heat treating was originally developed for the processing of electron-tube and space-age material. As the technology has progressed, vacuum heat treating has been found to be extremely useful in many less exotic metallurgical areas. Vacuum heat treating consists of various thermal treatments carried out in a heated chamber evacuated to a vacuum, or a partial pressure suitable for the particular material.

Vacuum heat treating can be used for:

1) Preventing reactions at the surface of the work, such as oxidation or decarburization, and retaining a clean surface
2) Removing surface contaminants from fabrication operations, such as oxide films and residual traces of lubricants, which are severe contaminants to the furnace
3) Adding a substance to the surface layers of the work, such as carbon, which results in carburization
4) Removing dissolved contaminants from metals, such as hydrogen or oxygen from titanium
5) Joining metals by brazing or diffusion bonding

Solution Treatment

Solution heat treatment consists of heating an alloy to a suitable temperature, holding that temperature long enough to allow one or more of its constituents to enter into solid solution in the alloy, and then cooling rapidly enough to hold the constituents in solution, leaving the alloy in a supersaturated, unstable state. Time at temperature should be kept to a minimum because excessive grain growth may cause an orange-peel surface or cracking in subsequent forming operations, as, for example, with austenitic stainless steels.

While solution heat treatment in a vacuum furnace does not necessarily require different heating and cooling cycles from those employed with any other type of furnace, use of a vacuum furnace offers the possibility of controlling the heating and cooling rates. However, when heat treating in vacuum, there are some other considerations that must be addressed in planning the solution treating cycle.

Loading Guides

One potential cause of failure in an annealing operation is lack of knowledge of the temperature distribution within the load. For many annealing applications, it is sufficient simply to specify that the steel be cooled in the furnace from a designated austenitizing temperature at a rate equivalent to air. But other applications may require rapid cooling to attain the physical properties desired. Thus, the application must be known and the furnace must be capable of controlling as well as documenting the cooling rate.

Fig 2.1 Workload stacked and placed on vacuum furnace grid for heat treatment

Typical cycles include the same precautions that would be inherent in any other batch type processes. It is important that the work load be distributed as uniformly as possible in the furnace to ensure that even heating and cooling as well as uniformity are maintained. Figure 2.1 shows a typical load of type 304 stainless steel parts ready for heat treatment. The recommended process cycle for this type of load is:

1) Pump down to 1 micron or less
2) Heat at a rate of 17 °C (30 °F)/min. to 650 °C (1200 °F)
3) Hold at temperature for 15 min.
4) Heat at a rate of 17 °C (30 °F)/min. to 870 °C (1600 °F)
5) Hold at temperature for 15 min.
6) Heat at a rate of 17 °C (30 °F)/min. to 1080 °C (1975 °F)
7) Hold at temperature for 1 h
8) Force-cool to 65 °C (150 °F)

As specified, holding at given temperature levels to allow the load to equalize to the furnace temperature is a good vacuum heat-treat practice.

Vacuum Levels

A frequently posed question concerns the vacuum level required for a given process. It must be remembered that the use of vacuum generates a physical reaction. It is a mistake to try to use it as a vacuum cleaner, or to rely on it to remove dirt, oil, or any other foreign matter from the workpiece. While the vacuum furnace will clean up the work to an extent, the contami-

Table 2.1 Vacuum levels for heat treating selected metals

Material	Operation	Pressure Range, torr
Tool Steel Series A, D, H, T, or M	Hardening	1×10^{-1} to 1×10^{-4}
Stainless Steel Ferritic, Martensitic, and Austenitic (stabilized and unstabilized)	Annealing, Hardening and Brazing	5×10^{-4} to 1×10^{-5}
Superalloys, Nickel or Cobalt Base	Solution Heat Treating, Aging and Brazing	1×10^{-4} to 1×10^{-5}
Titanium Alloys	Solution Heat Treating	5×10^{-4} to 1×10^{-5}
Titanium Alloys	Degassing	1×10^{-2} to 1×10^{-5}
Aluminum Alloys	Brazing	5×10^{-5} to 1×10^{-6}

nants will condense out in the cooler portions of the furnace and subsequently add to the outgassing requirements or contamination of subsequent loads. If all work going into the vacuum furnace were simply to be thoroughly pre-cleaned, its pumping system could be smaller and its maintenance problems greatly reduced with no loss in quality of the work produced. A typical vacuum level for the load shown above is 1 micron or less during the complete cycle. This will ensure that the vacuum furnace will produce bright workpieces. Table 2.1 lists vacuum levels for heat treating selected metals.

Heating Rates

Because the heating rate under vacuum is governed almost entirely by radiant energy, this can greatly affect the manner in which the work should be loaded. Workpieces with large mass and heavy sections heat at approximately the same rate as in an atmosphere furnace, and the rule of thumb of 1 h/in. of cross section applies. When the workpieces are randomly loaded, some can shield others from heat radiated by the heating elements and delay their heating. To overcome any such possible delays it may be necessary to rack workpieces of this type, or at the very least allow sufficient heating time to ensure that the heaviest mass section reaches temperature.

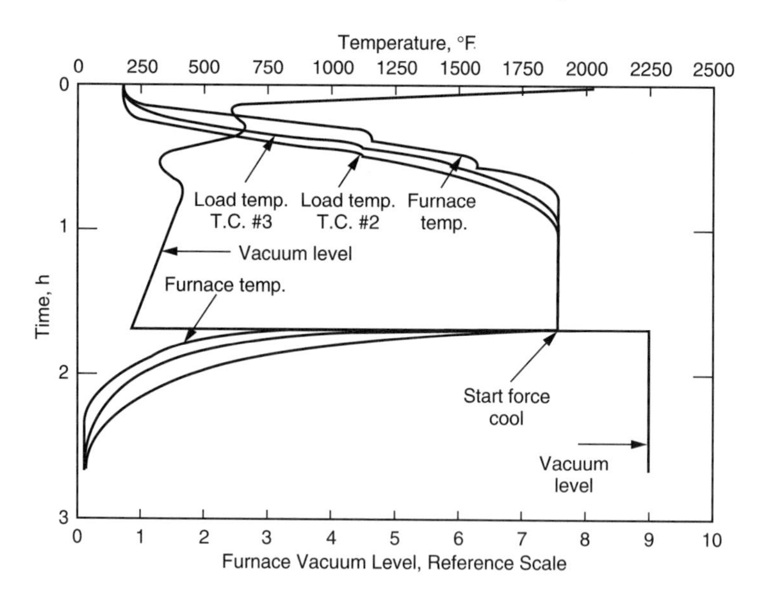

Fig 2.2 Typical vacuum furnace recorder chart showing traces of two load thermocouples and furnace controlling thermocouple

Because the load shown above has a relatively dense mass, to ensure that the inner portion of the load reaches temperature and holds it long enough to acquire the metallurgical characteristics needed, a minimum of two thermocouples should be placed in it to monitor the center portion of the work. Because of the radiation heating effect, the temperature at the center of the load lags behind that read by the furnace thermocouple, and the outside of the load heats faster than the inside (Fig 2.2). Thermal soaking at 650 °C (1200 °F) and 870 °C (1600 °F) allow the load heat levels to stabilize as the furnace heats up to temperature, to ensure even treatment of the workpieces and help reduce thermal shock and distortion.

Outgassing

Outgassing is the spontaneous evolution of gas from a material in a vacuum; its source is gas trapped in tight crevices, blind holes, porous metal, or other discontinuities. During normal heat cycles, any contaminants in the system generally manifest themselves as outgassing, which is detectable by increases in the heating chamber internal pressure level. As the furnace stabilizes and its heating rate slows or stops, volatilization of the contaminants equalizes and the furnace attains the desired vacuum levels.

Leak-Up Rate

The leak-up rate of a vacuum furnace is determined with the chamber empty, at a pressure less than 5×10^{-4} torr (0.5 microns) and isolated from the effects of the pumping system. After pump-down, with the pumping system disengaged, the pressure increase per unit of time is its "Leak-Up Rate." Normal leak-up rate levels are in the range of 10 to 20 microns/h.

Cooling Methods and Rates

There are three basic means of cooling the workload of a vacuum furnace:

1) Vacuum cooling,
2) Static cooling, and
3) Forced cooling.

Vacuum cooling consists of turning off the power at the end of the heating part of the cycle and allowing the load to cool at the same rate as the furnace without any outside means of forced cooling. Under these conditions, the load cools at a slow rate because of the virtual absence of convective heat transfer in the vacuum.

Static cooling consists of backfilling the furnace with a dry gas following the heating portion of the cycle, to promote convective transfer of heat from the work to the water-cooled shell. This will enhance the heat transfer of the vacuum-cool cycle, but will still yield a relatively slow, uncontrolled cooling cycle.

Forced cooling is the most time-efficient approach to furnace use. After the heating part of the cycle, the furnace is backfilled with gas to a pressure less than atmospheric, and a circulating fan moves the gas over the workload and over the water-cooled coils of a heat exchanger, to assist its convective circulation. Argon, nitrogen, helium, or hydrogen are generally used as backfill gases and quenching agents in vacuum furnace systems. Table 2.2 lists typical analyses of these gases as used in this type of service.

The three methods used to increase the rate of heat removal are:

1) Increasing the mass flow of the gas (i.e., gas velocity),
2) Increasing gas pressure, and
3) Using a gas with a higher thermal conductivity.

Partial-Pressure Systems

Among the many reasons for using partial pressure in a vacuum furnace application are (1) to prevent vaporization of some alloys, (2) to help prevent pressure bonding, and (3) to reduce the possibility of eutectic melting. The partial-pressure system is a means by which a furnace can be operated at a pressure other than the maximum pressure achieved by its pumping system. This is accomplished by metering a controlled amount of gas into the furnace chamber to hold it at the desired pressure. The partial-pressure

Table 2.2 Typical analyses of backfill gases

Gas	Purity, %	Impurity, ppm					Dew Point F	Cooling Rate[a]
		O_2	N_2	CO_2	CO	H_2		
Argon	99.9995	2	2	1	...	1	–110	.074
Nitrogen	99.9993	3	...	...	...	...	–110	1.00
Helium	99.998	1	10	...	...	1	–80	1.03
Hydrogen	99.9	10	1500	1	2	...	–75	1.40

(a) Relative to nitrogen as 1.0

system is generally operated at any setting between 10 and 1000 microns. Operation procedures include pumping down the furnace to a lower pressure to eliminate residual air at a temperature where vaporization is not a problem and then bleeding the partial-pressure gas into the furnace. The diffusion pump is closed, the roughing pump is kept running, and the internal pressure is automatically held constant by the partial-pressure control system.

Temperature Monitoring

During the vacuum cycle it is important to know the temperature variations of the load as well as the furnace temperature itself. Typically, the furnace thermocouple is inserted through the furnace wall along with an over-temperature thermocouple. In addition, a furnace load generally has at least one load thermocouple, although it is not uncommon to use as many as nine for any given load, depending on the recording and monitoring requirements of the application. When only one is used, it is usually placed in the center of the load, at the coldest location, to keep the operator informed as to load temperature in relation to furnace temperature. These thermocouples provide temperature information to the furnace control microprocessor. Quite often furnace cycles must be adjusted to ensure the load is heating at the same rate as the furnace and will conform to the cycle required for the specific application.

Annealing

Annealing is a process by which a distorted cold-worked structure reverts to a stress-free state through the application of heat. This stress-free structure promotes the required mechanical properties, which in turn ensure dimensional stability of the material and enhance its machinability and formability. The annealing process is governed by many of the same precautions that apply to normalizing. Both the heat-up and cool-down rates in the vacuum furnace are typical annealing processes in any other types of furnace. With modern furnace equipment, heat-up and cool-down rates of the load are readily and easily controlled and maintained through this portion of the cycle by the furnace control microprocessor.

As with most vacuum processes, annealing loads must be properly set up. To ensure uniform heating and cooling rates, the workload should be arranged in the furnace in a way that will maintain as constant and uniform a mass as possible. Figure 2.3 shows a typical load set up for vacuum annealing, with the workpiece placed directly on the furnace grid. The process cycle for this type of load is:

Fig 2.3 Aircraft parts on grid ready for processing

1) Pump down to 1 micron or less
2) Heat at a rate of 17 °C (30 °F)/min. to 650 °C (1200 °F)
3) Hold at temperature for 30 min.
4) Heat at a rate of 17 °C (30 °F)/min. to 870 °C (1600 °F)
5) Hold at temperature for 30 min.
6) Heat at a rate of 11 °C (20 °F)/min. to 1150 °C (2100 °F)
7) Hold at temperature for 1 h
8) Vacuum cool to 870 °C (1600 °F) at a rate of 22 °C (40 °F)/h
9) Backfill with argon and force-cool to 65 °C (150 °F)

The vacuum level must be maintained at a minimum of 0.1 micron throughout the entire vacuum heat-up cycle. The controlled-cooling portion of the cycle, at the rate of 22 °C (40 °F)/h is programmed to customer specifications. When the cycle reaches this point, the furnace is backfilled with argon to complete the cycle. In this application the load temperature is monitored by four load thermocouples (not shown in the photo) located at its outside corners.

Fig 2.4 Typical annealing load of stainless steel drawn reservoirs

Another example of a typical annealing load consists of stainless steel drawn reservoirs, as shown in Fig 2.4. The workpieces are stacked inside a stainless steel grid to contain them and utilize the full available volume of the furnace. Stainless steel wire mesh screen is placed between the work-pieces to prevent them from bonding together during processing. A typical cycle for this type of load is:

1) Pump down to 1 micron or less
2) Heat at a rate of 17 °C (30 °F)/min. to 650 °C (1200 °F)
3) Hold at temperature for 15 min.
4) Heat at a rate of 17 °C (30 °F)/min. to 870 °C (1600 °F)
5) Hold at temperature for 15 min.
6) Heat at a rate of 17 °C (30 °F)/min. to 1080 °C (1975 °F)
7) Hold at temperature for 15 min.
8) Force cool to 65 °C (150 °F)

Vacuum levels of 1 micron or less for such loads will produce bright workpieces that will meet the metallurgical requirements of the customer.

Hardening

Hardening changes the metallurgical structure of a load through the heating and cooling process. When hardening is done in a vacuum furnace, special consideration must be given to placement and spacing of the loads. In the case of small workpieces, whether in baskets or on the grid, load density must be considered. Both the heat-up and cool-down rates will affect the metallurgical result of the hardening cycle, and they can produce characteristics that are not wanted. If these rates are not uniform throughout the load, distortion will occur in large workpieces, and too slow a cooling rate may result in low hardness.

A uniform heating rate is critical, for example, for the surface of an H-13 tool steel injection molding die, because it will minimize grain-boundary carbide formation, which will cause heat checking. The vacuum furnace has simplified the heat treating of tool steel. Because the heating of large tool sections can be closely controlled, distortion and warpage is minimized. Therefore, because subsequent grinding, grit-blasting, and scale removal are not required, basic tool sizes and shapes can be machined to near-finished

Fig 2.5 Typical vacuum heat treatment load of assorted air-hardening tool steel workpieces

conditions. Cooling rates vary, depending on the type of tool steel being heat treated, and experimentation using the basic knowledge that is available will help eliminate warpage problems. The vacuum technology of today has the capability of positive pressure quenching from 2 bar to 20 bar (15 psi to 220 psi), with 2 bar to 6 bar most common in the United States. This gives greater control over cooling, allowing for faster cooling or for cooling at a specific rate.

In vacuum hardening, proper thermocouple placement will assure that all workpieces attain programmed temperatures. Typical placement puts a thermocouple in both the thinnest and the heaviest sections, where the two serve as temperature controllers.

Vacuum levels must be controlled for the hardening of high-speed tool steel grades in order to prevent vaporization of surface constituents, particularly chromium and manganese, which will affect hardenability of the material. However, the workload can be raised to preheat temperature under a high enough vacuum to degas the surface.

Figure 2.5 shows a typical heat-treat load of mixed air-hardening tool steel parts. Note the special loading arrangement, for equal spacing and

Fig 2.6 Typical load of M-2 high-speed tool steel

density distribution, to ensure uniform heat-up and cool-down rates. The process cycle for this type of load is:

1) Pump down to 1 micron or less
2) Heat at a rate of 20 °C (35 °F)/min. to 650 °C (1200 °F)
3) Hold at temperature for 45 min.
4) Heat at a rate of 20 °C (35 °F)/min. to 815 °C (1500 °F)
5) Hold at temperature for 30 min.
6) Heat at a rate of 20 °C (35 °F)/min. to 1025 °C (1880 °F)
7) Hold at temperature for 1 h
8) Force cool to 65 °C (150 °F)

Techniques for cooling tool steel in the vacuum heat-treating process vary substantially, depending on the metallurgical characteristics required for the end-use application. If desired, the heat-treating process cycle above could include vacuum tempering without removing the load from the furnace. By programming continuous cycles, the furnace can be made to produce finished tool steel workpieces that have been hardened and tempered under vacuum. While the vacuum tempering process is a costly one, at times it offers the most economical alternative.

Figure 2.6 shows a typical load of high-speed M-2 tool steel. The process cycle for this type of load is:

1) Pump down to 1 micron or less
2) Heat at a rate of 20 °C (35 °F)/min. to 675 °C (1250 °F)
3) Hold at temperature for 30 min.
4) Heat at a rate of 20 °C (35 °F)/min. to 845 °C (1550 °F)
5) Hold at temperature for 30 min.
6) Heat at a rate of 20 °C (35 °F)/min. to 1160 °C (2125 °F)
7) Hold at temperature for 15 min.
8) Force cool to 65 °C (150 °F)

Tempering and Aging

Although vacuum furnaces can be used for the tempering and aging processes, heat transfer in the vacuum furnace occurs by radiation. Aging and tempering cycles are performed under approximately 540 °C (1000 °F), at which temperature true radiation from the furnace elements is inefficient, and therefore, temperature uniformity is less consistent. However, the introduction of partial pressure gas into the vacuum furnace will have a slight convection effect and this will help to offset the lack of heat radiation. The circulating gas will help keep the temperature evenly dispersed throughout the vacuum chamber and, therefore, help in the heat transfer.

One problem encountered in vacuum tempering and aging cycles is that of the cleanliness of the resultant work. Because it is not practical to process under extremely low vacuum pressures, the quality of the work is not always the bright luster finish that it was before the process.

When using partial pressures of hydrogen, even at extremely low dew points, the process is very inconsistent, resulting in varying degrees of brightness. Water vapor and oxygen in the hydrogen gas and the insulation of the furnace make this operation difficult.

Brazing

Brazing is defined as the joining of metals by a flowing thin layer, of capillary thickness, of nonferrous filler metal into the space between the metals. Bonding results from the intimate contact produced by the dissolution of a small amount of base metal and the molten filler metal, without fusion of the base metal. Sometimes the filler metal is put in place as a thin solid sheet or as a clad layer, and the composite is heated, as in furnace brazing.

Vacuums are especially suited for brazing very large, continuous areas where (1) solid or liquid fluxes cannot be adequately removed from the interfaces of brazing, and (2) gaseous atmospheres are not completely efficient because of their inability to purge occluded gases evolved at close-fitting brazing interfaces. Vacuum is suitable for the fluxless brazing of similar and dissimilar base metal combinations such as molybdenum, niobium (columbium), tantalum, titanium, and zirconium. The characteristics of these metals are such that even very small amounts of atmospheric gases may result in embrittlement and sometimes disintegration at brazing temperatures. These metals and their alloys may also be brazed in inert gas atmospheres if the gases are of sufficiently high purity to avoid contamination. It is interesting to note, however, that a vacuum system evacuated to 10^{-5} torr contains only 1.3×10^{-6}% residual gases based on a starting pressure of 1 atm (760 torr).

Vacuum brazing has the following advantages and disadvantages compared to other high-purity brazing atmospheres:

- Essentially, vacuum removes all gases from the brazing area, thereby eliminating the necessity to purify the supplied atmosphere. Vacuum brazing is generally done at pressures varying from 0.5 torr to 500 millitorr and above, depending on the materials brazed, the filler material used, the area of the brazing interfaces, and the degree to which gases are expelled by the base metals during the braze cycle.
- In vacuum, certain oxides of the base metal will dissociate at the brazing temperature. Vacuum is used widely to braze stainless steel, superalloys, aluminum alloys, and, with special techniques, refractory materials.
- In vacuum brazing, the difficulties sometimes experienced with contamination of brazing interfaces due to base metal expulsion of gases are negligible. Occluded gases are removed from the interfaces immediately upon evolution from the base metal.
- The low pressure existing around the base and filler metals at elevated temperatures removes volatile impurities and gases from the metals. Frequently the properties of the base metals themselves are improved. Because of the low surrounding temperatures, this characteristic is also a disadvantage where the filler, base metal, or elements of them, volatilize at brazing temperatures. This tendency, however, can be corrected with proper brazing techniques, or the use of a partial pressure gas.

The typical precautions that apply to any other brazing process are also relevant to vacuum brazing. Assemblies to be brazed should be placed in the furnace so they do not overlap, providing the most constant and uniform mass possible. This will enhance the uniformity of the heating and cooling rates.

A typical actual load set up for vacuum nickel brazing of one of the many nickel-based brazing alloys is shown in Fig 2.7. Note that the workpieces are placed directly on the furnace grid in a fixture that holds them vertically throughout the braze operation. Also note the insulating cloth under the braze fixture. This cloth prevents the braze alloy from falling onto the elements or furnace bottom. The process cycle for this type of load is:

1) Pump down to 1 micron or less
2) Heat rate of 17 °C (30 °F)/min. to 650 °C (1200 °F)

3) Hold at temperature for 20 min.
4) Heat rate of 17 °C (30 °F)/min. to 845 °C (1550 °F)
5) Hold at temperature for 30 min.
6) Heat rate of 11 °C (20 °F)/min. to 1085 °C (1985 °F)
7) Hold at temperature for 15 min.
8) Force cool to 65 °C (150 °F)

A vacuum level of 0.1 micron must be maintained throughout the entire vacuum heat-up cycle. The hold times may be increased to compensate for the braze alloy, or binder burn off, or outgassing. The additional hold time will ensure the proper vacuum level for brazing. Once the cycle reaches the force-cool portion of the program, the furnace is backfilled with argon to complete the force-cooling portion of the cycle.

The workload shown in Fig 2.8 consists of nickel-brazed assemblies which have completed the process cycle. Note their clean appearance and their evenly spaced arrangement on an insulating cloth placed on the furnace grid to protect it from damage by the braze alloy. Load temperature was monitored by two thermocouples (not shown in photo) on its outside corners.

Fig 2.7 Stems and cups ready for vacuum nickel brazing

Fig 2.8 Nickel-brazed assemblies

Ion Nitriding

Diffusion processing in a vacuum, specifically ion nitriding, is a relatively new process and most industries are only just beginning to explore its use. For more than 60 years, gas nitriding has been used as a means of imparting wear resistance and improved fatigue strength with a minimum of distortion. With advances in equipment, ion nitriding, sometimes called plasma nitriding, is becoming a very dependable and cost-effective process. It is a process used to surface harden metals for a wide variety of applications. As its name implies, this process uses ionized nitrogen gas, alone or in combination with other gases, to react with the work surface. Ionized nitrogen produces a characteristic purple glow around the workpieces while they are being treated.

This process requires:

- A vacuum vessel to remove possibly contaminating gases such as air
- An 800 to 1000 vdc power supply for the plasma discharge through which nitrogen ions are accelerated to the workpiece surface

- A gas distribution system to provide proper mix ratios and flow rates, and
- A pressure control system to maintain pressures of 1 to 10 torr, appropriate for supporting plasma discharge within the desired voltage range

In operation, the dc potential is placed across the hearth plate on which the workpiece rests and the vessel itself, with the hearth plate as the cathode (negative potential) and the vessel as the anode (positive potential). The workpieces shown in Fig 2.9 are typical of those that are ion nitrided for increased wear resistance. The inner and outer surfaces of these pieces are the working surfaces of the final assembly, and the increased wear resistance imparted to this prehardened type 4140 alloy achieves its final application requirements. The furnace cycle for this type of heat treatment is:

1) Pump down to 1 micron or less
2) Heat at a rate of 17 °C (30 °F)/min. to 525 °C (975 °F)
3) Hold at temperature for 24 h
4) Force cool to 65 °C (150 °F)

A typical gas mixture for this kind of processing is 25% nitrogen and 75% hydrogen and the plasma discharge watt density is 0.007 amps.

Why is ion nitriding used? Typically, to impart wear resistance. The advantages of ion nitriding include:

- Potentially reduced cycle times for nitriding steel
- Reduced distortion
- Possibility of minimizing or eliminating finish grinding
- Improved metallurgical properties, often with lower cost materials
- Eliminates copperplate masking through use of simple mechanical masks
- Eliminates the "white layer" of conventional nitriding and imparts hard wear-resistant surfaces without brittleness, spalling, or galling
- Yields a uniform case on complex geometries
- Reduces operating costs
- Eliminates environmental problems caused by toxic salts and gases
- Precisely repeatable cycles reduce scrap

Fig 2.9 Ion-nitrided workpieces

Vacuum Carburizing

The usual method of gas carburizing in conventional equipment requires a gas-tight furnace to enclose the work to be processed, into which is fed a carrier gas from an external endothermic gas generator. The proper quantity of carburizing gas, such as methane or natural gas, is added to the carrier gas to supply the carbon needed for the carburizing reaction. Elaborate methods of controlling the carbon potential of the furnace atmosphere during the thermal cycle are necessary to ensure the desired composition of the carburized case.

If a vacuum furnace is employed, the process can be simplified and speeded up. The work to be processed is placed in the cold furnace and the pressure is reduced to approximately 10^{-1} torr. The furnace is heated to 1040 °C (1900 °F) to outgas the work, and the temperature is stabilized. Purified methane is then backfilled to approximately 50 torr and circulated over the work until the entire quantity of carbon necessary to produce the finished case has been absorbed. This results in a thin carbon case on the work, with surface carbon concentration approaching saturated austenite at 1040 °C (1900 °F), at approximately 1.7% carbon. The internal furnace pressure is now reduced to about 10^{-1} torr, and the carbon is allowed to diffuse inward. The slope of the originally steep carbon gradient flattens out and approaches that of conventional cases. No carbon is lost or added to the surface of the work during the diffusion period.

At the end of the diffusion cycle, the work may be furnace-cooled and then hardened in an operation using another furnace. In furnaces equipped for liquid quenching, the work can be furnace-cooled to a suitable quenching temperature and liquid-quenched in one continuous vacuum operation.

Some advantages of vacuum carburizing are:

- Higher furnace temperatures without reducing furnace life
- Elimination of endothermic atmosphere gas generators
- Vacuum furnace requires no gas or power during the idle period, and no conditioning period to attain proper carburizing conditions
- Faster carburizing rate because the surface carbon concentration is approximately 1.7% C instead of the more usual 1.1% C for normal gas carburizing
- The diffusion rate of carbon in iron is much higher at 1040 °C (1900 °F) compared with 925 °C (1700 °F), by as much as a factor of 4
- Shorter heating times offer an economic advantage, which becomes greater the deeper the desired case
- The work surface is outgassed and thus is better able to accept carbon

Because molybdenum furnace components, such as heat shields, heating elements, and work trays, are not harmed by the presence of methane at the usual vacuum carburizing temperatures, graphite heating elements are not mandatory for vacuum carburizing furnaces, although they are a popular choice for furnaces used for this purpose.

New Technology

The newest and most modern technology offers vacuum furnaces that are gas-fired instead of electrically powered. This affords the ability to process low-temperature vacuum cycles with a temperature uniformity of ±2.8 °C (±5 °F). In the gas-fired vacuum furnace, the work is processed in an enclosed chamber, and the circulating gas plenum is outside this chamber. The vacuum pumping system is similar to that of an electrically heated vacuum furnace, and the furnace has the capability of processing using partial pressures of argon, nitrogen, and hydrogen.

A typical process cycle for tempering or age-hardening in a gas-fired vacuum furnace is:

1) Pump the vacuum chamber to 1 torr
2) Backfill to 200 torr with hydrogen
3) Temper or age-process per application specifications
4) Force cool at the end of the cycle by blowing air through the outside of the plenum while maintaining the vacuum/partial pressure in the plenum
5) After room temperature is reached, purge the work chamber with nitrogen and open the furnace.

Selected References

1) *Ion Nitriding*, T. Spalvins, Ed., ASM International, 1987
2) George A. Roberts and Robert A. Cary, *Tool Steels*, 4th Edition, American Society for Metals, 1980

• 3 •

Vacuum Considerations for Heat Treatment

Robert F. Gunow, Sr., APT Consulting
Robert F. Gunow, Jr., Vac-Met, Inc.
Alexander J. Gunow, Integrated Technologies

Introduction

The behavior of metals and non-metallic materials in vacuum depends on the residual atmosphere, chemical properties, physical properties, vacuum levels, and temperature. These diverse influences on material behavior are further complicated by interactions that occur among them. However, a thorough analysis of the limitations and interactions of these factors can lead to greater understanding and appreciation of the vacuum processing environment. Anticipating the effect of these factors is a prerequisite to consistent quality processing of metal and metal alloys. This principle is relevant to quenching with gas or liquid, vacuum carburizing, ion nitriding, ion carburizing, and partial pressure processing. A review of the behavior of selected metals in vacuum processes will help describe the application of these factors and define the advantages and disadvantages of using vacuum vs other atmospheres.

Residual Gas and Vacuum Integrity

Even at very low pressure levels of 1×10^{-6} torr, the current accepted capability limit of production vacuum heat treating equipment, a cloud of gaseous molecules remains in the vacuum chamber. Analysis of this residual gas during pumping indicates that water vapor is one of its major constituents. For this reason, we must accept "vacuum" as a relative term and determine the effect of the partial pressure of water on the work at various vacuum levels and temperatures. Residual gases are particularly important when reactive metals are involved.

Residual gases are not only influenced by vacuum levels, but also by chemical and physical properties of materials and contaminants, and by the mechanical design of the equipment. Equipment leaks may be overcome by high-capacity pumps and the flow of a leaking atmosphere may contribute to unbalancing the expected equilibrium behavior. Such unexpected behavior also occurs with internally contaminated equipment, where the contaminant continues to vaporize during the heating portion of the cycle.

Chemical and Physical Properties

Much of the reference data is derived from pure-metal investigations, and such data does not always reflect the behavior of alloys which may contain complex intermetallic compounds and metal solid solutions. The presence of complex surface oxides in the vacuum environment also contributes to deviations from expected reactions for pure metals and metal oxides.

Many metal oxides, such as those found with common ferrous alloys, are dissociated during vacuum treatments, e.g., the dissociation of iron oxide (FeO) to iron and oxygen. As shown in Fig 3.1, iron oxide will tend to dissociate under conditions represented to the right of the FeO curve. For example, it will be reduced at a temperature of 260 °C (500 °F) and a vacuum level of 1 torr. In actual practice, a more rapid dissociation of the oxides of common ferrous alloys occurs in the vicinity of 980 °C (1800 °F) at 1 micron. Although the information in Fig 3.1 cannot be used to precisely predict reactions in a vacuum heat treat furnace, its information can be used to forecast the general direction a reaction will take. When dissociation takes place, a "cleaning action" takes place, resulting in a brighter and cleaner product.

Contaminants may be carried into furnaces from sources such as the workload and dirty baskets or fixtures. These contaminants may have a direct bearing on the vacuum levels that can be achieved, and/or they can become associated with the chemical reactions that take place during processing. Contaminants with a higher vapor pressure than the scheduled vacuum level must be removed by the pumping system prior to the final treatment. This removal usually occurs prior to any temperature reactions. The benign contaminants typically condense in the cooler areas of the furnace outside of the hot zone. Chemically active contaminants may react with the workload or furnace hot-zone materials. Other contaminants may dissociate into products that are either removed by the vacuum pump system or become inert with respect to subsequent processing.

Contaminants remaining in the furnace after processing may or may not be chemically harmful in subsequent processing. Particular attention should be given this possibility if subsequent processing involves higher tempera-

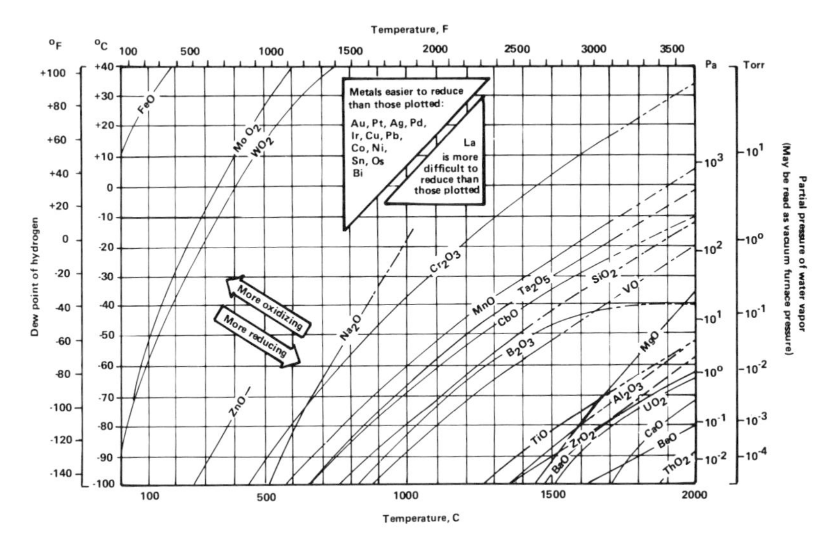

Fig 3.1 Metal-metal oxide equilibria in pure hydrogen atmospheres

tures than that which deposited the contaminants. Harmful contaminants include condensed vapors from volatile metals, volatilized oxides, organic compounds, and salts.

One way to minimize the effect of the remaining contaminants is to perform a periodic furnace cleaning cycle at 56 to 112 °C (100 to 200 °F) above normal operating temperatures, holding at temperature until vacuum levels approach the pressure capability level of the equipment. Caution should be exercised to ascertain that nothing is in the hot zone that will be damaged by the higher temperature.

Significance of Vapor Pressure

Vacuum heat treatment of cadmium, lead, zinc, other high-vapor-pressure elements, and alloys containing these elements, can result in sufficient evaporation of these constituents to affect surface integrity and in some situations alter the chemical composition of the base alloy. Table 3.1 shows the correlation between vapor pressure and temperature. As seen in the table, zinc will vaporize out of brass, a copper-zinc alloy, at a temperature of 250 °C (478 °F) and vacuum level of 10^{-4} mmHg (0.1 micron). After treatment under these conditions, a brass workpiece would become a porous copper-enriched residue. Copper evaporates at pressures less than 10^{-4} mmHg above 1035 °C (1895 °F).

Table 3.1 Vapor pressure and temperature correlation for metals

Symbol	Element	10^{-4} mmHg 0.1 micron	10^{-3} mmHg 1.0 micron	10^{-2} mmHg 10 microns	10^{-1} mmHg 100 microns	760 mmHg 760,000 microns
Al	Aluminum	1486 °F	1632 °F	1825 °F	2053 °F	3733 °F
Sb	Antimony	977	1103	1251	1434	2624
As	Arsenic	...	428	...	590	1130
Ba	Barium	1011	1157	1321	1524	2557
Be	Beryllium	1884	2066	2275	2543	...
Bi	Bismuth	997	1128	1290	1328	2588
B	Boron	2084	2262	2471	2712	...
Cd	Cadmium	356	428	507	610	1409
Ca	Calcium	865	982	1121	1292	2709
C	Carbon	4150	4480	4858	5299	8721
Ce	Cerium	1996	2174	2381	2622	...
Cs	Cesium	165	230	307	405	1274
Cr	Chromium	1818	1994	2201	2448	4500
Co	Cobalt	2484	2721	3000	3331	...
Cb	Columbium	4721	4602	...	...	...
Cu	Copper	1895	2086	2323	2610	5003
Ga	Gallium	1578	1769	1999	2278	...
Ge	Germanium	1825	2034	2284	2590	...
Au	Gold	2174	2401	2669	2995	5425
In	Indium	1375	1544	1746	1990	...
Ir	Iridium	3909	4244	4633	5092	...
Fe	Iron	2183	2390	2637	2916	4955
La	Lanthanum	2057	2268	2518	2820	...
Pb	Lead	1018	1148	1324	1508	3171
Li	Lithium	711	822	957	1125	2502
Mg	Magnesium	628	716	829	959	2025
Mn	Manganese	1456	1612	1796	1868	3904
Mo	Molybdenum	3803	4163	4591	5448	10056

(continued)

Table 3.1 Vapor pressure and temperature correlation for metals

Symbol	Element	10^{-4} mmHg 0.1 micron	10^{-3} mmHg 1.0 micron	10^{-2} mmHg 10 microns	10^{-1} mmHg 100 microns	760 mmHg 760,000 microns
Ni	Nickel	2295	2500	2750	3054	4950
Nb	Niobium	4271	4602	...	...	...
Os	Osmium	4107	4444	4833	5288	...
Pd	Palladium	2320	2561	2851	3198	...
Pt	Platinum	3171	3459	3794	4159	7965
K	Potassium	253	322	405	509	1189
Rh	Rhodium	3299	3580	3900	4274	...
Rb	Rubidium	190	253	329	423	1254
Ru	Ruthenium	3736	4046	4408	4831	...
Sc	Scandium	2122	2340	2593	2903	...
Si	Silicon	2041	2233	2449	2705	4149
Ag	Silver	1558	1688	1917	2120	4014
Na	Sodium	383	460	556	673	1638
Sr	Strontium	775	887	1020	1182	2523
Ta	Tantalum	4710	5108	...	...	...
Tl	Thallium	862	932	1123	1220	2655
Th	Thorium	3328	3630	3985	4408	...
Sn	Tin	1692	1850	2172	2318	4118
Ti	Titanium	2280	2523	2815	3168	...
W	Tungsten	5013	5461	5988	...	10701
U	Uranium	2885	3146	3448	3808	...
V	Vanadium	2887	3137	3430	3774	...
Y	Yttrium	2484	2721	3000	3331	...
Zn	Zinc	478	554	649	761	1665
Zr	Zirconium	3020	3301	3623	4014	...

Note that the vapor pressure of metals is fixed with probable values at a given temperature, and that the temperature at which the solid is in equilibrium with its own vapor descends as the pressure to which it is exposed descends. For example, iron must be heated to 4955 °F at atmosphere before its vapor pressure is greater than atmosphere (760 mm Hg); but, as shown here, this point is reached at 2390 °F at 1.0 micron.

Evaporation of metals can be minimized by slightly increasing the pressure in the vacuum chamber, by injecting a pure non-reactive gas such as argon, helium, or nitrogen to produce a chamber pressure that exceeds the vapor pressure of the metal. Silver, for example, may be exposed to temperatures as high as 1045 °C (1917 °F) without fear of evaporation if pressure within the chamber exceeds 10^{-1} mmHg. Gases with a minimum purity of 99.99% and dew point of –60 °C (–80 °F) or lower are usually used for this purpose.

Vacuum heat treatment of highly volatile cadmium, lead, zinc, and alloys of these metals also poses a threat to furnaces with internal heating elements, such as the popular cold-wall types. Conductive and condensed vapors near power leads and terminals can cause electrical shorts and arcing damage across these insulated conductors.

Temperature Considerations

Not only is vapor pressure dependent on temperature, but chemical reactions also are influenced by temperature in the vacuum environment, and this can result in the dissociation of compounds or the reversal of chemical reactions that normally take place at room temperature. Therefore, the choice of equipment material becomes important for specific operations.

Perhaps one of the most important thermal considerations in vacuum processing is solid-state diffusion, the movement of a constituent in a solid from an area of higher concentration to one of lower concentration, or the movement of atoms and molecules to new lattice sites. This action is stimulated by intimate contact between two reactive surfaces at elevated temperatures. If sufficient diffusion takes place, the surfaces adhere, producing surprisingly strong bonds. This effect is utilized in sintering powdered metal (P/M) workpieces.

High temperatures, the clean environment of vacuum furnaces, and forces that induce intimate contact can bring about unwanted sticking. Loading the workpieces so as to produce heavy pressure on those on the bottom can cause diffusion bonding. Differential expansion between two materials can also cause intimate contact. For example, a close-fit stainless steel plug inside tightly coiled plain carbon steel strip can exert pressure on the coils and cause sticking. One way to prevent this is to separate the sticking-prone surfaces with a non-reactive, stable, low-vapor-pressure material. Compounds used for this purpose are the oxides of aluminum, magnesium, titanium, yttrium, and zirconium, as well as nitrides of titanium and boron. These compounds may be applied to sticking-prone surfaces of fixtures, baskets, and workpieces, by thermal spraying, cold spraying, chemical vapor deposition, and painting. In brazing technology, these materials are known as "stop-off" agents and are readily available from suppliers of high-temperature brazing alloys. When heat treating sensitive metals, it is beneficial to bake coated items prior to vacuum treatments, in order to remove all potentially reactive volatile constituents.

Refractory Metals

Refractory metals are used where resistance to chemical attack is required or where very high temperatures are expected. Molybdenum, niobium, tantalum, and tungsten have very high melting points, which makes them candidates for furnace heating elements. Tantalum is frequently used for chemical vessels, while niobium has been employed for high-temperature rocket applications. Refractory metals will, however, react with common gases under certain conditions.

The refractory metals hafnium, molybdenum, niobium, tantalum, and tungsten, from Groups IV and V of the Periodic Table, are reactive at elevated temperatures with atmospheres containing carbon compounds, hydrogen, oxygen, and nitrogen. Figure 3.2 shows the results of oxygen attacking a niobium alloy after a protective coating has been penetrated.

Surfaces of tungsten and molybdenum are degraded by oxidation at stress-relieving and annealing temperatures. Figure 3.3 shows the effect on a molybdenum P/M alloy of short-term exposure to oxidation at high temperatures. Similarly, niobium and tantalum will oxidize in the presence

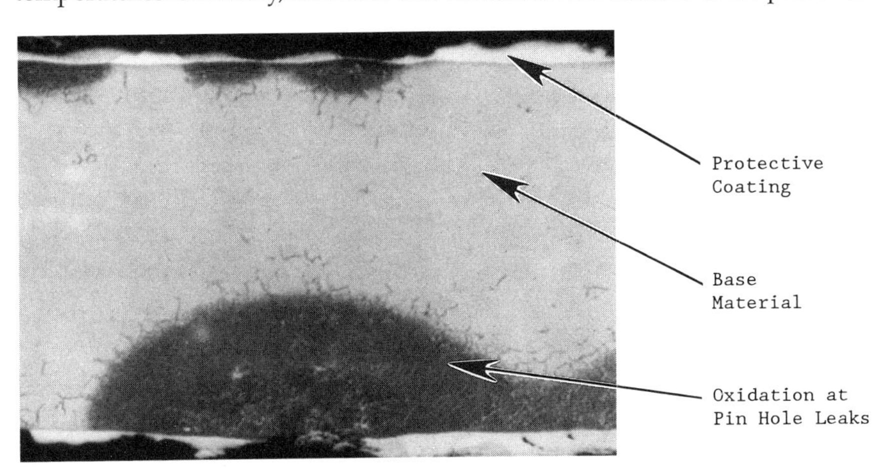

Fig 3.2 Oxidation of a niobium-5% titanium alloy at pin hole leaks in protective coating after 100 h in still air at 1150 °C (2100 °F)

Fig 3.3 P/M molybdenum part exposed to oxidizing atmosphere for 10 min. at 1040 °C (1900 °F), showing both surface and grain boundary oxidation. Electrolytic etch with 10% oxalic acid, 400×

of oxygen at temperatures above 300 °C (570 °F). Therefore a vacuum of at least 1×10^{-5} torr must be achieved prior to processing these materials above this temperature. In addition, these metals will absorb hydrogen and nitrogen at processing temperatures above 650 °C (1200 °F), with loss of desirable physical properties. Accordingly, an appropriate quench medium must be anticipated if cooling more rapid than vacuum cooling is required. Inert gases such as argon and helium can serve this purpose.

Hafnium has gas-phase reactions similar to zirconium, which will react at elevated temperatures with hydrogen, oxygen, and nitrogen to form oxides and hydrides. Zirconium hydride will dissociate in vacuum, and in fact the dissociation of zirconium hydride at vacuum levels below 1 micron and temperatures above 900 °C (1650 °F) has been used to produce zirconium powder.

Because refractory metals readily form carbides, atmospheres containing carbon should not be used for heat treating them. Vacuum requirements and maximum leak rates are particularly stringent for thermal treatment of refractory metals: pressure levels should typically not exceed 1×10^{-4} torr during the temperature cycles, and cold-leak rates should not exceed 2 microns/h.

Superalloys

Superalloys, used for high-stress applications at temperatures above 650 °C (1200 °F), include cobalt, iron, and nickel-base alloys. Table 3.2 lists some superalloys in each of these categories. Superalloys contain one or more of the alloying elements aluminum, chromium, cobalt, iron, molybdenum, nickel, niobium, titanium, and tungsten. Of these, only chromium and aluminum have relatively high vapor pressures at heat treating temperatures. Because of the small quantity of aluminum in these alloys and strong binding energies within the alloys, aluminum does not evaporate with any rapidity. Binding energies are attractive forces between atoms of a given crystalline structure, which inhibit the escape of surface atoms to a vacuum atmosphere.

Chromium, however, evaporates noticeably at temperatures and pressures within normal heat treatment ranges. At a temperature of 990 °C (1818 °F) and a pressure of 1×10^{-4} mmHg, chromium will vaporize, and evidence of condensed chromium vapor can be found in the cooler areas of the furnace, usually in the form of rough, silvery nodules or a gray, frosty coating. In a leaky furnace, these deposits may be green in color. Long soaking times, such as occur in homogenizing treatments of castings, and high temperatures, such as are required for solution treatment of direction-

Table 3.2 Nominal compositions of selected superalloys

Designation	Cr	Ni	Co	Mo	Ti	Cb+Ta	W	B	Al	V	Fe
Iron Base											
A286	15	25.5		12	2.1			0.006		0.3	Rem.
Incoloy 800	21	32.5			0.38				0.38		45
19-9DL	19.5	9.5		1.4	0.22		1.4				Rem.
Nickel Base											
Inconel 718	19	52.5		3	0.9	5.1			0.5		18
René 41	19	53	11	9.8	3.2			0.006	1.6		
M252	19	54	10	10	2.6			0.006	1		
Cobalt Base											
L605	20	10	52				15				
Haynes 188	22	22	40				14.5				La-.0
Elgiloy	20	15	40	7							16

ally solidified and single-crystal castings, tend to promote the loss of chromium. One solution to this problem is to inject an inert gas into the furnace, to increase the total pressure above the partial pressure of chromium at the processing temperature. Some specifications, such as General Electric C50TF94, which includes treatment of René 142, allow pressure up to 400 microns to deter the depletion of alloying elements.

The alloying elements aluminum and titanium are frequently added to superalloys to induce precipitation hardening and the formation of tight, temperature-resistant oxides. For subsequent operations after heat treating, it is imperative to inhibit the formation of these surface oxides in a vacuum furnace. This is particularly important because aluminum and titanium are sensitive to minute amounts of oxygen and water vapor at heat treating temperatures, and the oxides may form during thermal processing from small partial pressures of residual gases in the vacuum chamber or contaminants in a quenching gas. When aggregate amounts of 1.5% Al and Ti are formulated in alloys, the resulting alloy surfaces are sensitive to oxidation. The surface oxides are various shades of gray or blue. Low leak rates of 5 microns/h and vacuum levels below 0.1 micron are recommended to avoid the formation of these oxides.

Required cooling rates for solution treatments or superalloys can be attained using gas- and liquid-quenching vacuum furnaces. Cooling rates of 28 °C (50 °F)/min. may be attained with standard gas-quenching equipment. However, workloads and loading geometry must be compatible with a particular furnace design. Quenching rates of 42 °C (75 °F)/min. or more are accomplished in 5- and 6-bar or high-velocity gas quenching furnaces. Oil quenching is rarely employed for superalloys, but it remains an alternative.

Argon gas at 99.99% purity, with a dew point below 25 °C (80 °F), is often used for gas quenching. Helium has been used to accelerate quenching rates, and nitrogen will also produce a faster cooling rate than argon, but some aircraft specifications do not allow the use of nitrogen for high-temperature treatments. Figure 3.4 compares the relative cooling properties of quench gases.

Stainless Steels

Stainless steels are classified as ferritic, martensitic, austenitic, and precipitation-hardening grades. Table 3.3 shows some examples of these classifications. All stainless steels are routinely processed in vacuum furnaces. The martensitic and precipitation-hardening stainless steels can be hardened, stress-relieved, and annealed. The ferritic and austenitic grades are non-hardenable and can only be stress relieved and annealed.

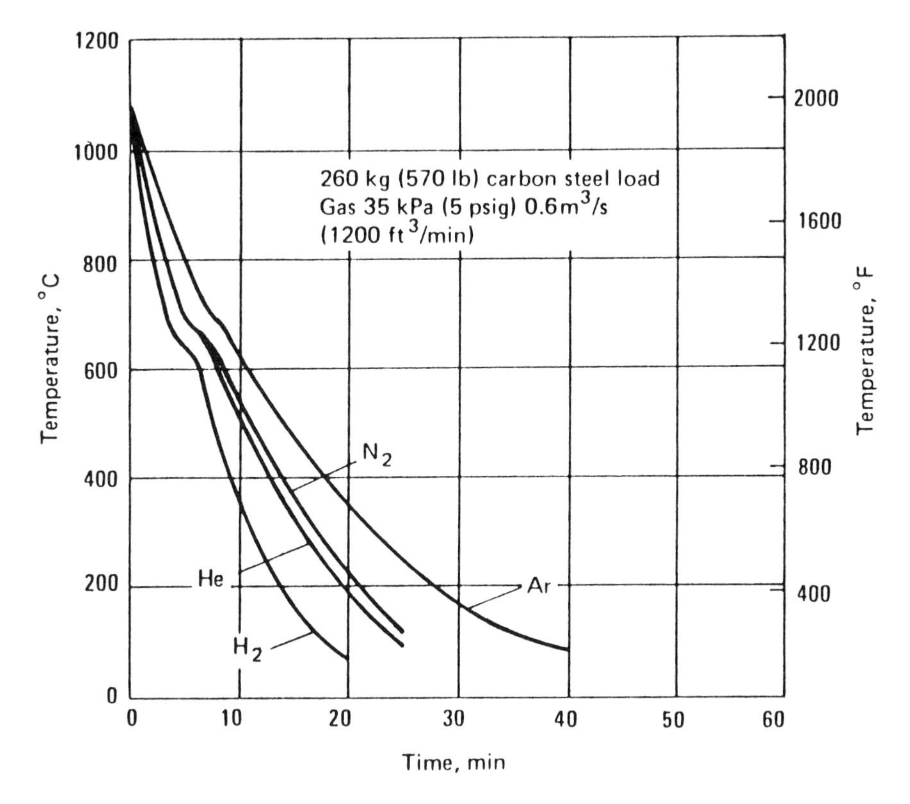

Fig 3.4 Quench gas effectiveness

Table 3.3 Classification of some common stainless steels

Ferritic	405, 409, 430 and 434
Martensitic	403, 410, 420 and 440C
Austenitic	302, 304, 321 and 347
Precipitation-hardening	17-4ph, 17-7ph, AM350 and PH 15-7 Mo

Surface brightness, and/or freedom from carburization and decarburization is usually a requirement for parts made from these steels. When brightness is not a factor, lower pressure levels are not required and furnaces without diffusion pumps may be used. To assure products that are free from heavy oxidation and decarburization, vacuum levels should be 2×10^{-2} torr or lower and the vacuum chamber should have a leak-up rate of less than 20 microns/h.

Surface discoloration, ranging from tan to blue, resulting from treatment in the temperature range of 595 to 730 °C (1100 to 1350 °F) can be a problem with stainless steels. The same materials at higher temperatures and pressures remain bright. Although this is a complicated matter from a thermodynamics viewpoint, some insight as to the cause of this phenomenon may be obtained from the oxidizing and reducing curves of Fig 3.1.

Formation of chromium oxide is the principle reason for stainless steel discoloration. Residual gas analysis has shown that in a cold chamber evacuated to less than 10^{-2} torr, water remains as the predominant constituent. The Cr_2O_3 curve shows that at 595 °C (1100 °F), the partial pressure of water vapor must be below 10^{-3} torr to provide a more reducing residual atmosphere and prevent oxidation. At 980 °C (1800 °F) the partial pressure is approximately 10^{-1} torr. Films of complex oxides can form on stainless steel at temperatures between 40 and 980 °C (100 and 1800 °F), and the pressure requirements to achieve an active reducing or dissociating residual atmosphere in this critical range of temperatures are much lower than 10^{-1} torr. Thus, clean and bright stainless steels will remain bright after treatments at 980 °C (1800 °F) and vacuum levels of 10^{-2} torr, while the same material will be discolored after treatments at 595 °C (1100 °F), even at the same vacuum levels. In practice, oxidation of the stainless steel takes place on the way up to 980 °C (1800 °F), and then at a temperature somewhat below that point the oxides are reduced or dissociated. Perhaps this reduction is related to the dissociation of water vapor in vacuum at elevated temperatures.

Introducing a partial pressure of a pure neutral gas such as argon, helium, or nitrogen, lowers the partial pressure of water vapor and aids in keeping surfaces bright. For this purpose, vacuum levels are generally adjusted to approximately 300 microns to 1×10^{-1} torr, while injecting a flow of neutral gas above 540 °C (1000 °F). The vacuum pumps can be throttled during the injection of neutral gas to reduce the flowthrough volume, which in turn reduces the cost of this procedure. Throttling is a particularly significant cost factor at higher partial pressures.

Gas quenching with standard vacuum furnaces usually provides a sufficient cooling rate to harden the martensitic grades of stainless steels and solution-treat the precipitation-hardening grades. However, within the specified chemistry range of 416 stainless steel, liquid quenching may be necessary to obtain the desired hardness. Section thicknesses over 13 mm (½ in.) with a specification hardness over 36 HRC may require quenching in a 5-bar or high-velocity gas quenching furnace.

When maximum protection from intergranular corrosion is required, annealing the unstabilized austenitic stainless steels may require a more rapid quench than can be attained by gas quenching, particularly for section thicknesses of over 13 mm (½ in.). Oil- or water-quenching may be necessary. Stabilized grades such as 321 and 347 or specified low-carbon grades such as 304L that avoid grain boundary carbides, may be gas quenched. The application of the end product usually determines the selected quenching medium. Compare Figs 3.5(a) and 3.5(b), which show grain boundary carbides present in air-cooled 304 stainless steel but absent in water-quenched 304 stainless steel.

Chromium depletion is usually not a problem with the process cycles normally used for heat treating stainless steels. Some surface chromium is depleted during high-temperature brazing operations, but because chromium is in solution in iron and the diffusion rates are low, the depletion is negligible.

Because the evaporation rate of metals depends on surface area as well as temperature, evaporated and condensed deposits are more evident when powdered metals are used in brazing and sintering operations. These deposits accumulate on the colder walls of the furnace and will continue to grow in thickness with each cycle. (See the section in this chapter on Superalloys.)

In processing stainless steels it is important to note that solid-state diffusion between nickel and carbon occurs at temperatures of approximately 1165 °C (2130 °F), which may cause eutectic melting. Caution should be exercised when nickel-bearing stainless steels are in contact with carbon fixtures above this temperature. Carbon will also solid state diffuse in contact with ferritic stainless steels at hardening temperatures. See Table 3.4 for other critical combinations.

Tool Steels

Tool steels are selected on the basis of strength and wear resistance, and after heat treatment their surface chemistry must remain free from carburization and decarburization. These steels are generally classified as air-hardening, oil-hardening, and water-hardening. Vacuum heat treating equipment may be used to process air- and oil-hardening tool steels. Current equipment limitations do not permit the general use of vacuum equipment for heat treating water-hardening tool steels. Although hardening temperatures for vacuum heat treatment are similar to those prescribed for salt and gas-atmosphere processing, procedures are somewhat different in that preheating is not necessary. Thermal shock is avoided because vacuum furnaces are relatively cold when loaded. Heating rates, soaking temperatures, and soaking time can be programmed to compensate for mass and cross-section thickness, which results in uniform and repeatable cycles.

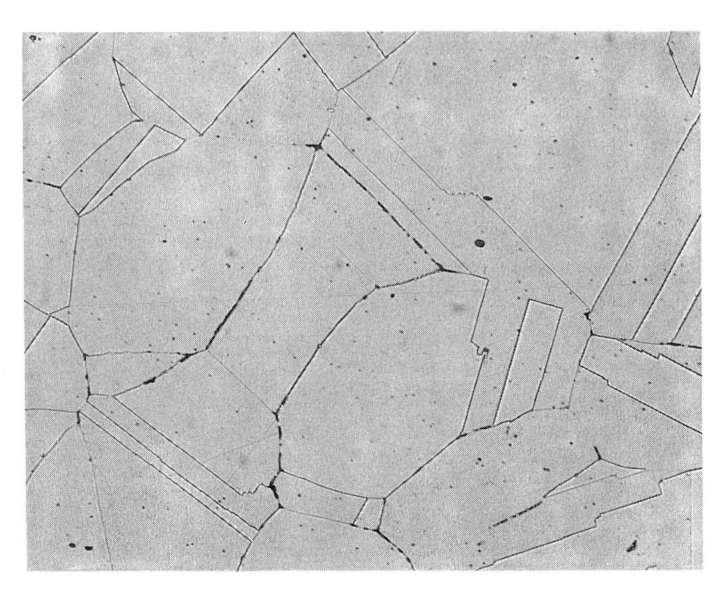

Fig 3.5(a) Type 304 stainless steel, air-cooled from 980 °C (1800 °F), with grain boundary carbides. Electrolytic 10% oxalic acid etch, 400×

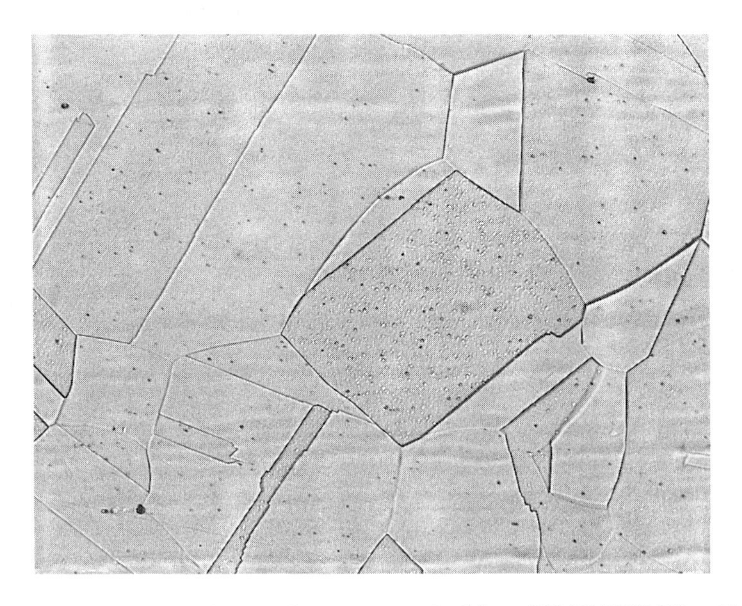

Fig 3.5(b) Type 304 stainless steel, water-quenched from 980 °C (1800 °F), without grain boundary carbides. Electrolytic 10% oxalic acid etch, 400×

Table 3.4 Eutectic combinations and temperatures at which solid-state diffusion can cause melting

Eutectic Combination	Temperature
Moly/Nickel	2310 °F (1265 °C)
Moly/Titanium	2210 °F (1210 °C)
Moly/Carbon	2700 °F (1480 °C)
Nickel/Carbon	2130 °F (1165 °C)
Nickel/Tantalum	2450 °F (1340 °C)
Nickel/Titanium	1730 °F (940 °C)
Titanium/Carbon	2600 °F (1425 °C)

Section thickness, dimensional requirements, and hardness specifications determine the selection of the quenching medium. Air-hardening steels, such as M-2, with a section thickness of 50 mm (2 in.) and a hardness requirement of 62 HRC may be gas quenched in a standard 2-bar vacuum furnace. Similarly, M-4 with the same section thickness will achieve a hardness of 64 HRC. D-2 and A-2 with 8-inch section thicknesses can be hardened to 58 to 60 HRC, while H-13 with the same cross section is normally limited to 50 to 54 HRC. These examples of air-hardening tool steel treatments do not necessarily express the limits of section thickness and hardness for these alloys. Larger section thicknesses for these materials with the same hardness requirements can be accommodated with 5- and 6-bar vacuum quenching systems as well as high-velocity gas systems. Surfaces of vacuum processed air-hardening tool steel are brighter and cleaner than those produced in conventional gas-atmosphere furnaces.

A typical automatic programmed cycle for hardening of M-2 steels is:

1) Load the work into a cold furnace, in baskets or fixtures, and position the workload thermocouples
2) Pump down chamber to 1 micron or less. (If the furnace is not diffusion-pump-equipped, pressure will be 10 to 20 microns)
3) Heat to 780 to 845 °C (1450 to 1550 °F) at the rate of 17 °C (30 °F)/min., which allows the material to heat slowly through transition
4) Hold at temperature for 10 to 15 min. (Add 5 min. for every inch over 3 in. of cross section)
5) Heat to 1095 °C (2000 °F) at the rate of 17 °C (30 °F)/min.
6) Hold at temperature for 10 to 15 min.
7) Heat to 1205 °C (2200 °F) at the rate of 17 °C (30 °F)/min.

8) Hold at temperature for 5 min.
9) Rapidly backfill chamber with the quenching gas and circulate from the hot zone to heat exchanger
10) Cool to 120 °C (250 °F), as indicated by workload thermocouples
11) Equalize furnace pressure and room pressure and remove load

Furnace loads should be restricted to weights and volumes that will ensure uniform heating and cooling and prevent prolonged heating above 1095 °C (2000 °F).

Pure dry nitrogen is normally used as a quenching gas. Helium and mixtures of helium and nitrogen may be used to produce more rapid gas cooling, but the cost is higher. (See Fig 3.4.)

The results of oil quenching in vacuum will yield the same or similar results as those obtained from oil quenching in an atmosphere furnace. Vacuum atmospheres are much less difficult to monitor and control than generated gas atmospheres, and they provide a more consistent quality.

Surface integrity resulting from vacuum heat treatments permits pre-machining of tools close to or at finished dimensions. Coated tools, such as those coated with titanium nitride by means of vapor deposition are manufactured with no additional grinding or machining after coating and heat treatment. Their dimensions, therefore, must be held to within 51 to 76×10^{-4} mm/mm (2 to 3×10^{-4} in./in.). Precise temperature control and controlled uniform cooling rates with modern vacuum gas quenching furnaces make it possible to achieve dimensions within these tolerances. Figures 3.6, 3.7, and 3.8 show representative samples of vacuum heat-treated tool steels.

Other Steels

The main reason for processing low-, medium-, high-carbon, and low-alloy steels in vacuum is to control surface reactions such as carburization, decarburization, oxidation, etc. This can be accomplished easily by using a relatively soft vacuum of approximately 25 microns or less in a clean vacuum furnace. This vacuum range will produce clean, bright work.

Four basic types of vacuum furnaces are used to process these steels:

Gas-Quenching

Equipment of this type is useful only for annealing, stress relieving, tempering and other non-hardening operations, because it cannot achieve the critical cooling rates necessary for heat-treat hardening. Because most of

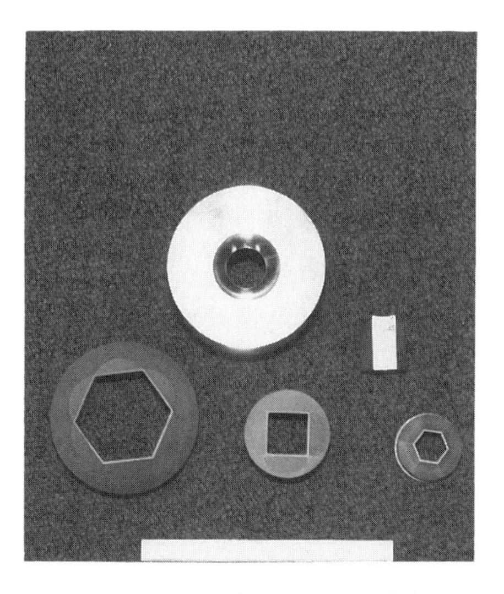

Fig 3.6 Tin-coated M-2 tool steel dies

Fig 3.7 Vacuum-hardened D-2 and M-2 tool steels after application of wear-resistant coatings. Darker surfaces coated with tungsten carbide

Fig 3.8 M-2 tool steels hardened and tempered in vacuum after coating

these processes are low-temperature operations and vacuum furnaces transfer heat by radiation, these steels can generally be processed in conventional atmosphere equipment at lower costs.

Liquid-Quenching

These furnaces have the combined advantages of a vacuum atmosphere during heating and the fast quench needed to harden many of these steels. The exceptions are those steels that require water-quenching to attain full hardness.

Carburizing

These units are used to carburize high-quality items such as gears and intricate shapes, and case-depth and carbon content can be accurately controlled with them. With the introduction of modern oxygen probes, use of conventional atmosphere furnaces is usually more economical.

Ion-Nitriding and Ion-Carburizing

Ion nitriding is growing rapidly. This process offers reliable control parameters for producing accurate case depths with or without white layer, with shorter cycle times and the ability to nitride a variety of alloys. However, its versatility can be limited by nonuniform loading and certain workpiece configurations. Ion nitriding of tool and dies after hardening and tempering is now a common practice.

Vacuum ion carburizing has similar control advantages to ion nitriding but has not been widely accepted because of the perceived cost of processing vs conventional-atmosphere carburizing.

Copper and Brass

Copper and most copper alloys can be stress relieved, annealed, and heat treated in vacuum to prevent oxidation. Exceptions are those alloys that contain high vapor pressure elements, such as cadmium, lead, zinc, and alloys requiring water-quenching. Brass, with its high zinc content, will dissociate and form zinc vapor under high-vacuum conditions. Its evaporation rate can be reduced by using a partial pressure of an inert gas. Table 3.1 should be consulted for specific processing temperatures.

Annealing brass, which contains a high percentage content of zinc, requires a minimum partial inert gas pressure of approximately 100 microns (0.1 torr). Note that in Table 3.1, zinc vapor pressure is 100 microns at 405 °C (761 °F). In practice, nitrogen is usually used for this purpose and in this example the partial pressure would be kept at a favorable 500 to 1000 microns (0.5 to 1 torr).

Precipitation-hardening copper alloys which require water quenching from solution temperature are better treated in atmosphere furnaces, where a quick transfer to the quenching bath is more easily accomplished. Subsequent aging or precipitation treatments can be performed in vacuum equipment.

Copper brazing in vacuum is normally carried out under partial pressure conditions above the vapor pressure of copper.

A typical vacuum brazing cycle is:

1) Load prepared assemblies in furnace
2) Pump down chamber to less than 1 micron
3) Heat to 870 °C (1600 °F) at a rate of 22 °C (40 °F)/min.
4) At 870 °C (1600 °F) backfill with nitrogen to a pressure of 1000 microns. Hold pressure until cooling backfill
5) Heat to 1010 °C (1850 °F) at a rate of 22 °C (40 °F)/min.
6) Hold at temperature for 20 min. (Sufficient time to equalize the temperature of assembled workpieces)
7) Heat to 1105 °C (2025 °F) at a rate of 11 °C (20 °F)/min.
8) Hold at temperature for 5 min.
9) Furnace-cool to 980 °C (1800 °F) by discontinuing heat input
10) Backfill with nitrogen and force-cool to 95 °C (200 °F)
11) Equalize furnace and room pressure
12) Unload furnace and inspect brazed joints

Aluminum

Aluminum and its alloys are naturally protected by a thin film of oxide that forms in air. Because dry air or liquid salts can be used as a heat treating medium for it, there is little or no advantage in using vacuum for these treatments. Further, since vacuum furnaces heat by radiation alone, it is difficult for them to maintain temperature uniformity of ±2.8 °C (±5 °F) in the range of 595 °C (1100 °F). The solution treatment of aluminum alloys is within this range, and it requires stringent uniformity. The need to water-quench aluminum alloys from solution temperatures also complicates the use of vacuum. Vacuum heat treating of aluminum simply is not an economical process at this time.

Fluxless brazing of aluminum in high-vacuum is a practical procedure. The use of molten salt baths containing fluoride salts provides both the medium for heating and the flux for brazing. The salt bath process has been the more common method of brazing aluminum. To prevent corrosion after brazing, it is essential to remove all the salt. Frequently, this is difficult, particularly in areas entrapped by assemblies.

Specialized vacuum furnaces are designed to accommodate vacuum brazing of aluminum alloys. Furnace interiors must be kept dry and clean, to prevent contamination of sensitive aluminum surfaces. Magnesium is used in the furnace as a degasser, to gather the residual oxygen and water vapor in the heating chamber, and to assist in breaking up oxide films. It may be added to the brazing alloy or placed in the furnace in a suitable container from which magnesium vapor is generated.

The brazing of aluminum radiators for the automotive industry is a good example of vacuum furnace brazing. These high production furnaces are multiple chamber units. Except for maintenance, the primary heating chamber is protected from exposure to air and is kept under vacuum conditions by continuous pumping. A typical brazing cycle is:

1) Open front outer door, place assemblies on fixtures in loading chamber, and close front outer door
2) Pump down loading chamber to 10^{-5} torr
3) Open front inner door to heating chamber and transfer load to preheating zone held at 10^{-4} to 10^{-6} torr
4) Close front inner door and bring loading chamber to room pressure
5) Preheat workload to 425 °C (800 °F)
6) Transfer load to brazing zone
7) Heat to 595 °C (1100 °F) and hold at this temperature for 2 min.
8) Open rear inner door and transfer load to cooling chamber held at 10^{-4} to 10^{-5} torr
9) Close rear inner door
10) Backfill cooling chamber and cool to 180 °C (300 °F)
11) Open rear cooling chamber door
12) Transfer load to exterior cooling stand, close rear outer door and pump down cooling chamber

Titanium

Because titanium and its alloys are very reactive at elevated temperatures, heat treating procedures must be carefully planned. These metals will absorb hydrogen from an atmosphere containing hydrogen, water vapor, or hydrocarbons. Pickling can also result in hydrogen absorption. At temperatures above 260 °C (500 °F) the formation of titanium hydrides is accelerated. Hydrogen can embrittle titanium and its alloys, causing cracking when cooled from processing temperatures. It has a devastating affect on fatigue properties of titanium and titanium alloys, and it is usually limited to below

100 ppm for highly-stressed components. It migrates readily in titanium. Solid-state degassing of titanium to reduce its hydrogen content can be carried out at 705 °C (1300 °F) with vacuum levels below 1 micron. Since it cannot be detected by visual inspection, hydrogen content must be ascertained by chemical analysis.

Above 480 °C (900 °F) both oxygen and water vapor will produce surface oxides, known as "alpha case," with titanium. Nitrogen is absorbed at a much slower rate than oxygen, but it will form titanium nitride. Virtually no alpha case is tolerated in highly-stressed components, such as turbine engine parts. Alpha case appears as bright white surface layers in micrographs, whereas heavily oxidized titanium appears as brown to blue discolorations. (See Fig 3.9.) Alpha case can be removed by pickling, machining, or other mechanical means.

At temperatures above 945 °C (1730 °F) under vacuum conditions, solid-state diffusion of titanium and nickel occurs, which can result in eutectic

melting. Therefore it is imperative to keep titanium and its alloys free from contact with nickel alloy baskets and fixtures during heat treatment at or above the diffusion temperature. See Table 3.4 for other potentially troublesome solid-state diffusion combinations.

Because titanium and its alloys rapidly lose tensile strength and hardness above 455 °C (850 °F), consideration must be given to fixtures and supports for processing it above this temperature. Plain carbon steel has been used for fixture material, but compensation for dimensional changes as the result of phase changes above 715 °C (1325 °F) must be considered, and other noneutectic fixture material may be preferred. However, eutectic-forming materials can sometimes be utilized if separated from the titanium by nonreactive coatings. Spray or brush coatings of stable aluminum oxide and zirconium oxide have been used for this purpose.

Gas quenching of titanium should be carried out with argon or helium purified to 99.99% or better, with dew points below –60 °C (–80 °F).

When water quenching is required to attain physical property requirements, air furnaces are generally preferred, to facilitate quenching. The air must be dry and combustion products of gas fired furnaces should never come into contact with titanium and titanium alloys. Air heat treatments are usually followed by a pickling operation to remove alpha case.

• 4 •

Furnaces and Equipment

James G. Conybear, Abar Ipsen

Vacuum furnaces may be classified a number of different ways, for example, by method of heat containment, configuration of the furnace, or method of work transport. In this respect they are similar to other types of furnaces. Heat is contained in the furnace either by cooling the outside walls with water, or by insulating them. Thus, furnaces are also classified as either cold-wall (water cooled) or hot-wall. In addition to insulation, there is sometimes an outer shield separated from the main furnace shell, with air or water circulated between to further cool the outer wall. The most common design in use today is the cold-wall furnace with a double outer wall which serves as a cooling water jacket.

The most common approach to describing furnaces is by configuration, either horizontal or vertical, referring to the axis of the furnace shell, and method of work transport, batch or continuous. In batch furnaces the workload is charged as a single unit, or batch. In continuous furnaces work is fed into the furnace from one end and discharged from the other. In reality this latter type should be called a semi-continuous or indexing furnace, because the workpieces are charged through some kind of loading chamber where the pressure is reduced to the vacuum processing range.

Within these broad classifications is a wide variety of designs that allow for further subclassification, such as car bottom, roller rail, pusher, and retort. These are similar categories to those for other furnaces, usually referring to some configuration detail, such as the type of work support.

The selection of a particular type of vacuum furnace is determined by three considerations: design and performance, production factors, and quality. These will be discussed later in the chapter, following a more detailed review of the design classes.

Elements of a Vacuum Furnace System

A vacuum furnace system consists of several elements, the heart of which is the furnace itself. To it must be added the work-handling equipment, which can be manual or automatic, cleaning equipment, work storage facilities, trays or baskets, and other auxiliaries such as tempering and testing equipment.

Starting with the core of the system, the furnace itself, any furnace can be described as a set of modules, as shown in Fig 4.1. In addition to the basic

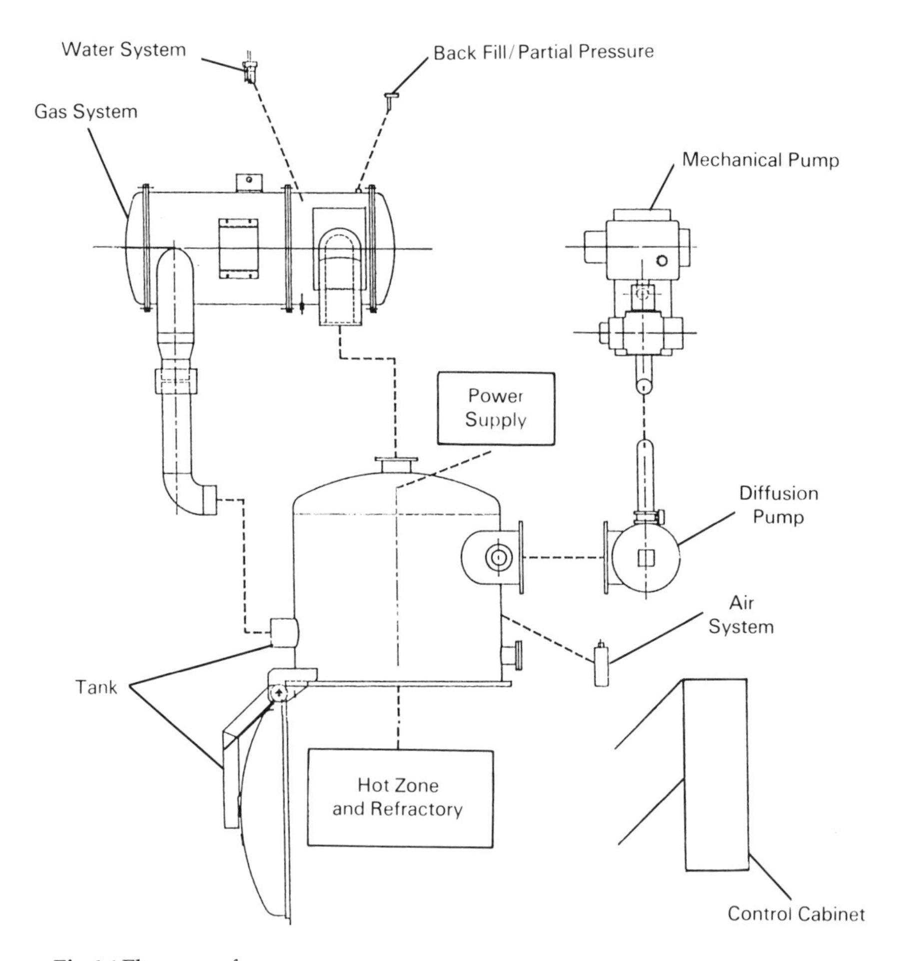

Fig 4.1 Elements of a vacuum system

shell, or tank, there are (1) a hot zone containing insulation or radiation shields, (2) heating elements, and (3) work supports. The other items shown are added to complete the total operating system. Some items that appear in the figure, such as the diffusion pump, may not be used. Others, such as the gas system, are frequently inside the furnace.

Hot-Wall Furnaces

The earliest vacuum furnaces were the hot-wall type. In this design, the entire vacuum vessel is heated by external means, either by electric elements or by burners. The heat is contained by insulation, similar to the way atmosphere chambers are constructed. Figure 4.2 shows this type of furnace in schematic form. Hot-wall furnaces, because of their large mass of insulating material, generally heat and cool slowly. They are also limited in temperature by the material of construction of the vessel, or retort, generally a heat-resistant alloy. Higher temperatures are sometimes achieved by pumping a vacuum on the outside of the vessel, as shown in the figure, which reduces stresses and eliminates oxidation of the retort, allowing the temperature to be increased closer to the melting point of the metal from which it is constructed.

Despite the limitations of this design, its simplicity and low cost keep it in some demand, although much less than the more common cold-wall furnaces. Figure 4.2 shows some of the additional features that can be incorporated. By extending the retort past the heated section, a water jacket can be added to accelerate cooling of the workload, or charge. The charge is loaded on a hearth or tray that can be moved in or out of the hot zone by a push rod extending through a seal in the outer end of the muffle. Such an

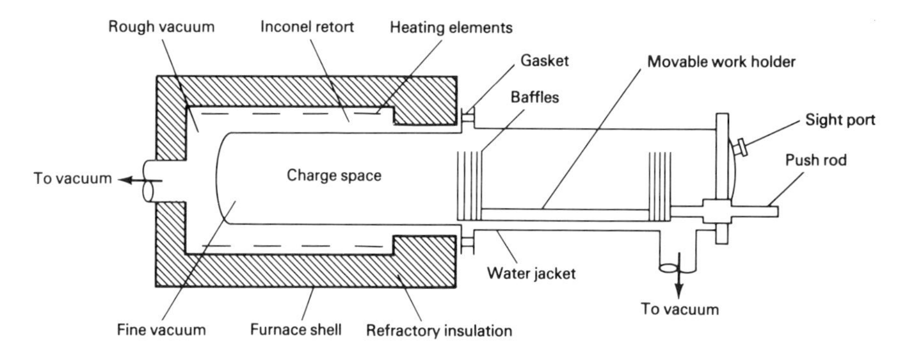

Fig 4.2 Hot-wall vacuum furnace with cooling section

approach can accommodate automatic transfer, inert, or even reactive atmospheres as a part of the process after vacuum-purging and inert-gas quenching using internal fans. However, this type of furnace is inherently slower and less flexible than the cold-wall design.

Cold-Wall Furnaces

Cold-wall furnaces consist of a water-cooled vacuum vessel with its exterior maintained near room temperature. As a result, the operating temperature is independent of the operating limits of the vessel. This permits very large vessels to be constructed of construction-grade alloy steels. Figure 4.3 shows the key features of this furnace type.

The water-cooled vessel is shown as a double-walled chamber, but can also be a single inner wall with cooling coils attached. Since there is a pressure differential of at least 1 atm between its inside and outside, the vessel is designed to pressure vessel standards. This water-cooled shell contains the heat insulation, the heating elements, and the hearth which supports the load.

Since the entire heating process is contained within the vacuum vessel, the vacuum acts as an insulating medium. This means that a very small mass of insulation is being heated relative to the workload. As a result, the furnace has low thermal inertia and can be heated and cooled very rapidly. On the other hand, because of the insulating effect of the vacuum, the workload gives up its heat slowly unless additional cooling is provided. Most modern furnace designs accomplish this by providing for backfilling the furnace with inert gas at the end of the treatment cycle and circulating this gas through a cooling path using blowers or fans. In recent years increased emphasis has been placed on increasing the rate of cooling by increasing the velocity or pressure, or both, of the backfill gas.

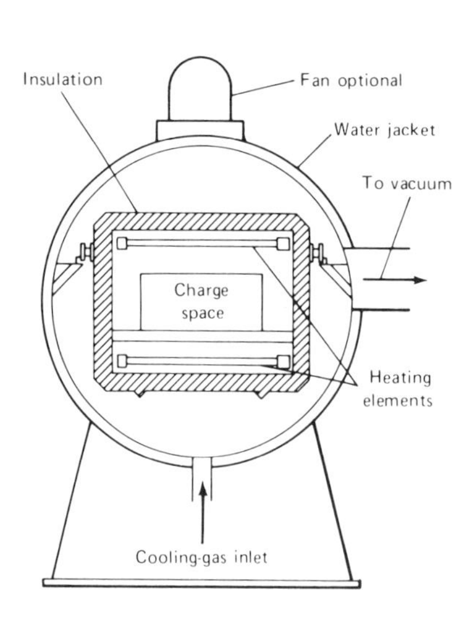

Fig 4.3 Cold-wall vacuum furnace

The development of the cold-wall furnace, with all heated parts contained within the protection of the vacuum space has permitted the use of materials

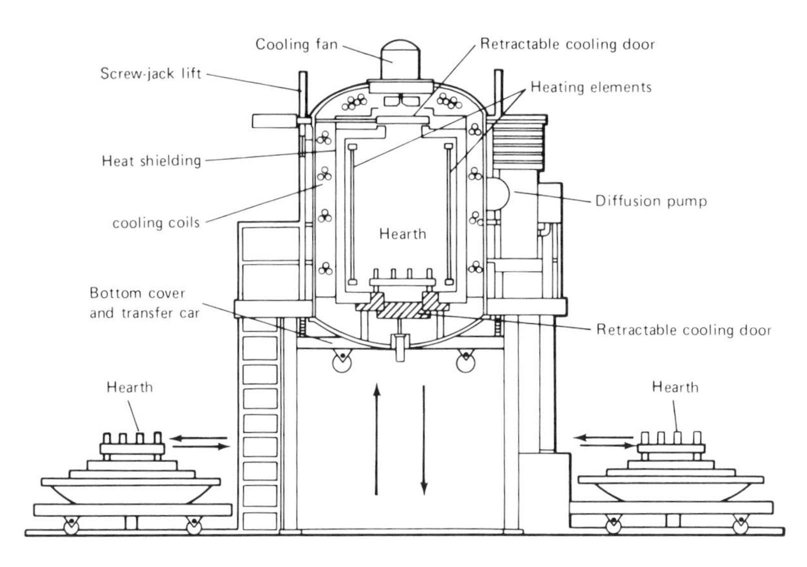

(a)

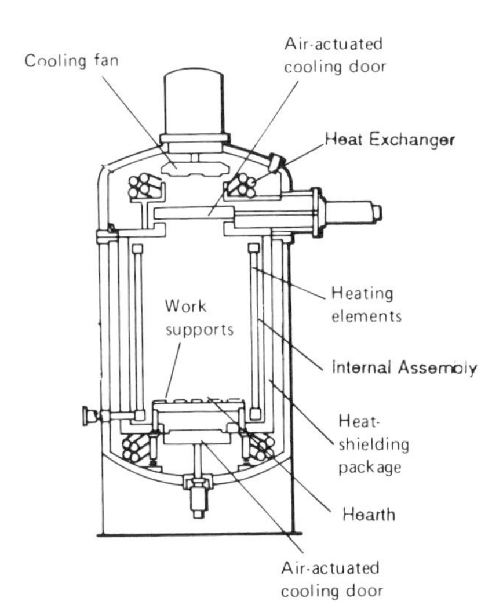

(b)

Fig 4.4 Batch vacuum furnace types. **(a)** Bottom load. **(b)** Pit type

that normally oxidize rapidly in even partial pressures of air, such as graphite, molybdenum, and tungsten, for insulation and heating elements. As these materials have become more available in many forms at reasonable cost, they have become the standard material of construction for these furnaces.

Batch Furnaces

The batch furnace may be horizontally loaded through a front door which is either side-hinged or opened vertically. Work is loaded directly into the heating section, which is at room temperature at the start of a cycle, and the workload remains stationary during heating.

The two other most common batch furnace configurations are the bottom- and top-loading furnaces, shown in Fig 4.4. The bottom-loading furnace is elevated, and when it opens the door moves vertically down on screws. The load is placed on this bottom head, which contains the work-support hearth. This design offers several advantages. These furnaces can carry very heavy loads, and the bottom head can be moved to one side for loading and instrumenting with thermocouples. The top-loading configuration is useful for processing long thin workpieces. It is loaded from the top, either by directly attaching fixtured work to the head or by hanging it from supports attached to the furnace wall.

As with atmosphere batch furnaces, the chief advantage offered by the batch configuration is its adaptability to a variety of cycles. Batch sizes can range from several to thousands of workpieces, depending on size. The capacity of the furnace is limited by its size, its design weight limit, and its power input. An additional factor, since the work is cooled in place, is the ability of its cooling system to remove heat from the load in an acceptable time period.

Batch Furnaces with Integral Quench. A variation of the batch furnace is the integral quench furnace, shown in Fig 4.5. In this type of furnace there

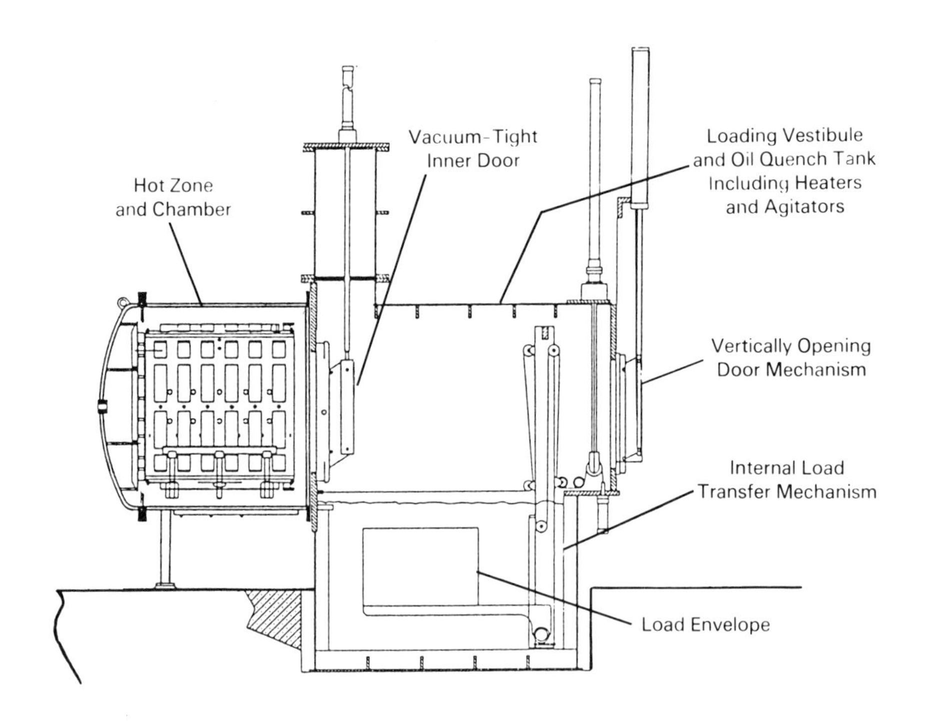

are two sections, one containing the hot zone as described above, and the other a quench medium, usually oil. Most commonly, these two sections are separated by a door with an automatic lift mechanism, although this may only be a non-sealing heat shield. The work charge is loaded on a transfer mechanism in the quench section and then transferred to the hot chamber under vacuum. This design offers the advantage of being able to harden oil-quenching grades of steel and hold the hot zone at temperature between cycles. It is classified as a batch furnace because the load is charged and discharged from the same end, one batch at a time. These furnaces are capable of processing loads up to about a ton.

Continuous Furnaces

Continuous furnaces (Fig 4.6) are used for large production loads. As mentioned above, this type is more appropriately called a semi-continuous furnace, because the load is indexed into it through some form of vacuum lock. However, it is continuous in that multiple charges are processed sequentially through it, from one end to the other.

These furnaces consist of at least three chambers. The first is a charging chamber, which may also be used for preheating the work. The second is the heating chamber, and some furnaces have multiple heating chambers. The third chamber is used for discharging the load, and it may also contain a gas-cooling system or quench. Continuous furnaces are equipped with an internal work carrier mechanism, for moving the load from one chamber to the next. Many varieties of these mechanisms have been built, including simple manual pushers, belt-type conveyors, walking beams, roller hearths, shaker hearths, and overhead monorail conveyors.

The key feature of these furnaces is that they are designed for higher production rates than batch units. As a result, they are usually equipped with more sophisticated controls for automatic operation. Their internal and external work carrier mechanisms, and door operators are interlocked and controlled automatically.

Furnace Hot Zone

Vacuum is in itself a very good insulator. By providing small vacuum spaces between the parts of a furnace system, heat flow from the inside to the outside of the furnace can be greatly reduced. Because of this, most vacuum furnaces have essentially a heated, insulated chamber within a chamber, held away from the cold wall by supports. A layer of insulation material is supported off this inner chamber, and it in turn encloses the heating elements, workload, and work supports.

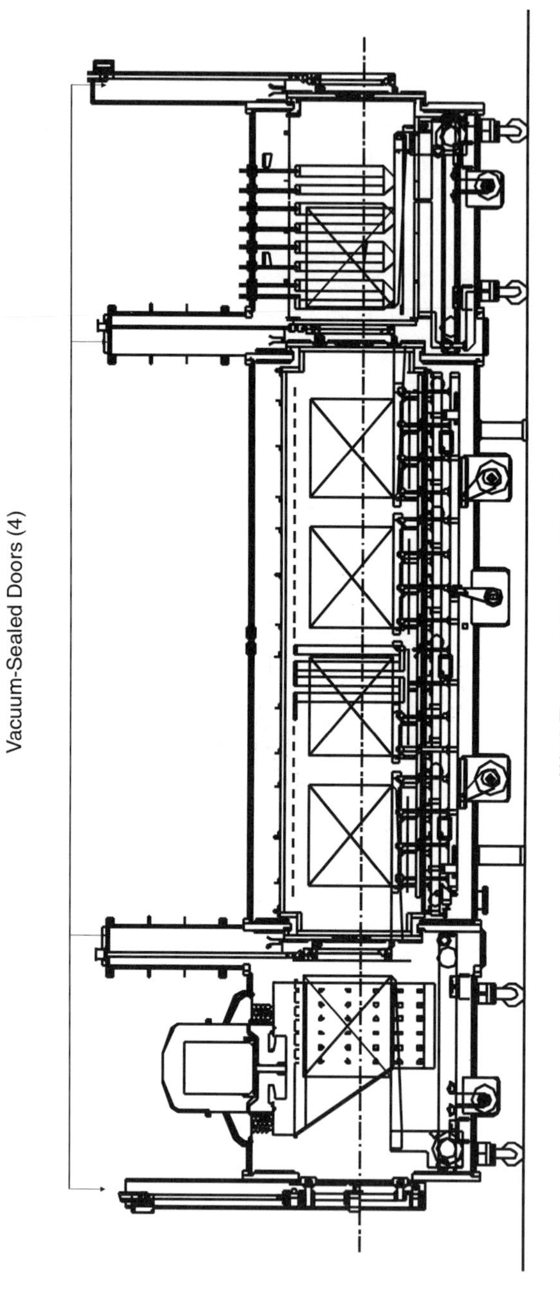

Fig 4.6 Continuous vacuum furnace

Two basic types of insulation have evolved, per Fig 4.7. The radiation shield type, shown in the top illustration, makes use of the fact that each time a vacuum space is provided between two adjacent parallel sheets of matter, the heat loss is reduced. Heat can transfer between the two only by radiation. The amount of heat transferred depends primarily on the temperature difference between them, and also on their surface quality. These shields can be any material required to resist the operating temperatures. Usually, the hot face and one or more inner shields are a refractory metal, or graphite for the highest temperatures, and the remaining shields are of a less expensive material, such as stainless steel.

The other common type, shown in the two lower illustrations, uses a hot pack of solid or fibrous insulation. Various combinations of metal and graphite or conventional insulating fiberboard or blanket are used. The insulating properties of blankets improve in vacuum because the spaces between the fibers are evacuated. The major disadvantage of such linings is the tendency of the fiber to adsorb moisture because of the large surface area of the fiber. This can be remedied by avoiding long periods with the furnace

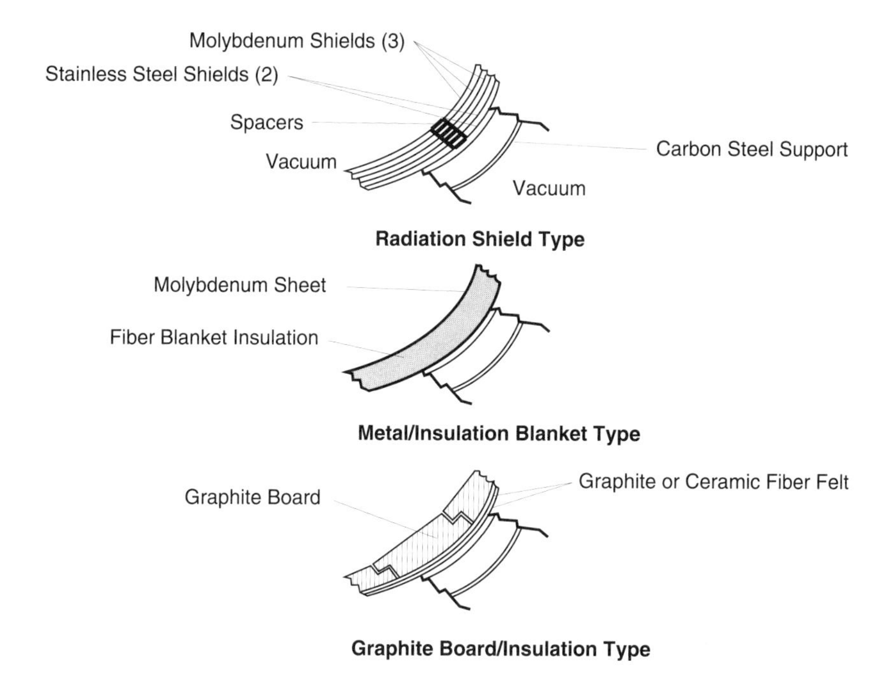

Fig 4.7 Vacuum insulation types

open. This type of insulation has been very successful, and offers the advantage of being easy to support.

Heating elements for vacuum furnaces are generally made of graphite, molybdenum, or tungsten, in sheet, rod, or tubular form. Because the vacuum protects the heating elements from oxidation, the high-temperature properties of these materials can be fully utilized. Many configurations of hot zones and heating elements have been used. The specific choice is a matter of preference, since all can be made to work equally well if good design and operating practice are followed.

Gas-Quenching Systems

To attain the cooling rate required for many alloys processed in vacuum furnaces, recirculated inert-gas cooling systems, such as those shown in Fig 4.8, are used. Both the external and internal cooling loops shown contain the same basic elements and operate similarly. A fan or blower recirculates an inert gas that has been backfilled into the system. This gas flows past the work through a duct, nozzle, or other opening into the hot zone. These openings may be closed off during the heating part of the cycle, to reduce radiation losses. After it exits the hot zone, the gas passes through a heat exchanger that is generally water-cooled.

The cooling capacity of a gas-quench is determined by the velocity, pressure, and properties of the gas. To achieve the cooling rate increases that have been necessary in each generation of vacuum furnace, the size of the heat exchangers and recirculating fan has progressively increased, and various types of cooling gas, including argon, helium, nitrogen, or gas blends, have been used.

In recent years pressures of over 2 bar have been used. As with velocity, for a given load, increasing the pressure increases the cooling rate due to its influence on the heat transfer coefficient. This relationship is shown generally in Fig 4.9. Note that there is a diminishing return with this effect, and the horsepower required to move the gas increases at the same time. For this reason the decision as to which particular combination to use is largely economic, although theoretically, continual increases are possible. Of course, once the backfill pressure exceeds atmospheric, the vessel must be designed for both negative and positive pressure. This means that an older furnace cannot simply be pressurized to improve its cooling rate.

Pumping Systems and Controls

The vacuum is obtained by various combinations of pumps, classified as roughing pumps and high-vacuum pumps. Mechanical pumps used in the initial stage of evacuation are simple positive-displacement pumps with

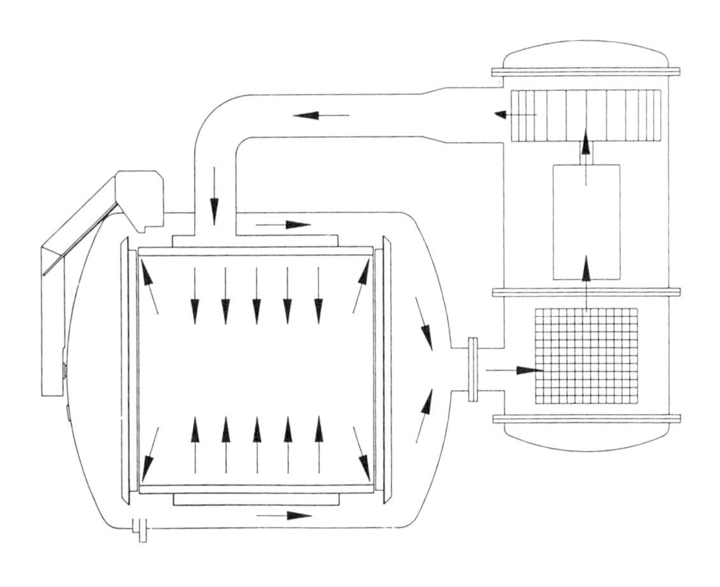

(a)

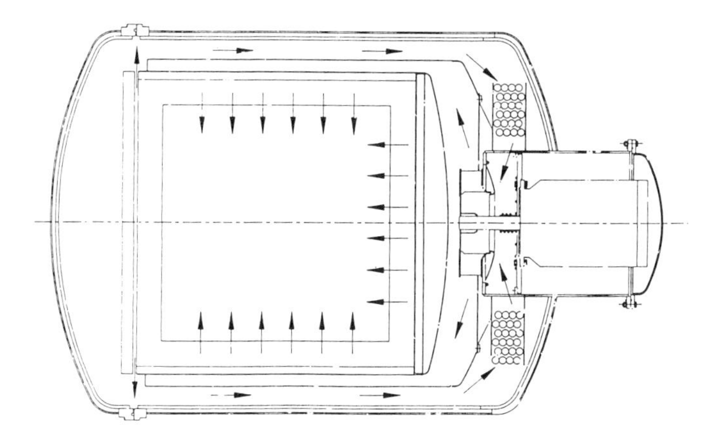

(b)

Fig 4.8 Vacuum cooling systems. **(a)** External blower and heat exchanger. **(b)** Internal fan and heat exchanger

suitable seals to operate in the low-pressure range. These are generally used with rotary blowers to improve pumping speed. This type of pump can

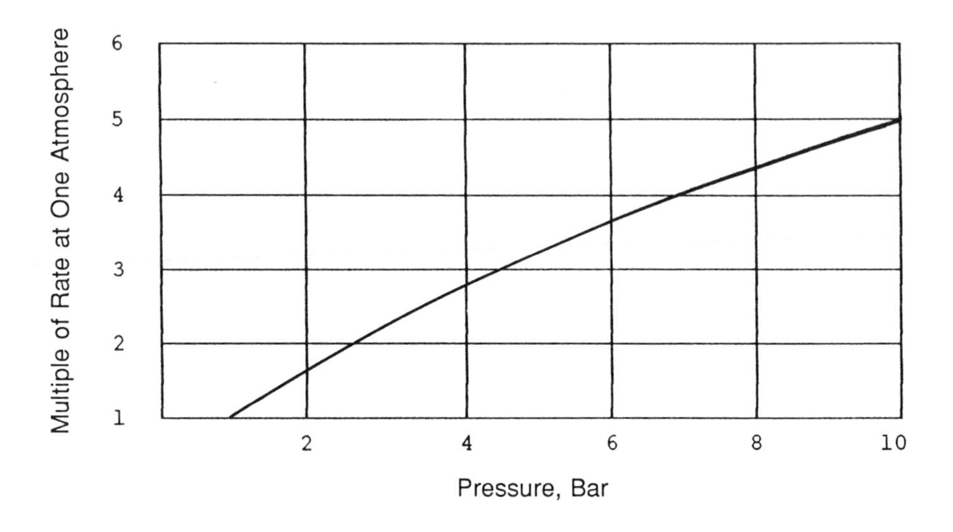

Fig 4.9 Influence of pressure on cooling rate

evacuate the furnace to between 1 and 20 microns. To obtain higher vacuum levels, vapor-diffusion pumps employing low-vapor-pressure oils or cryogenic fluids are used in combination with them. Pumps can be used in series, parallel, or combinations of both, to improve efficiency and speed. A typical system used on a high-purity batch furnace is shown in Fig 4.10.

Major advances have also been made in the control systems as well, to improve the flexibility of the furnaces. With modern process controls it is possible to provide controlled rates of heating and cooling, interrupted quenching, mechanical motion control, diagnostics, and statistical analysis. In this respect, vacuum furnaces are identical to their atmosphere counterparts, and they are often linked to such furnaces in integrated heat-treating lines.

Basic Selection Criteria

Per the foregoing discussion, there are many factors involved in the design and selection of a vacuum heat treat furnace. This has led to the wide variety of furnaces seen today, with many combinations of system elements. Considering production-related factors first, Table 4.1 summarizes the basic considerations to be addressed in choosing between a batch and a continu-

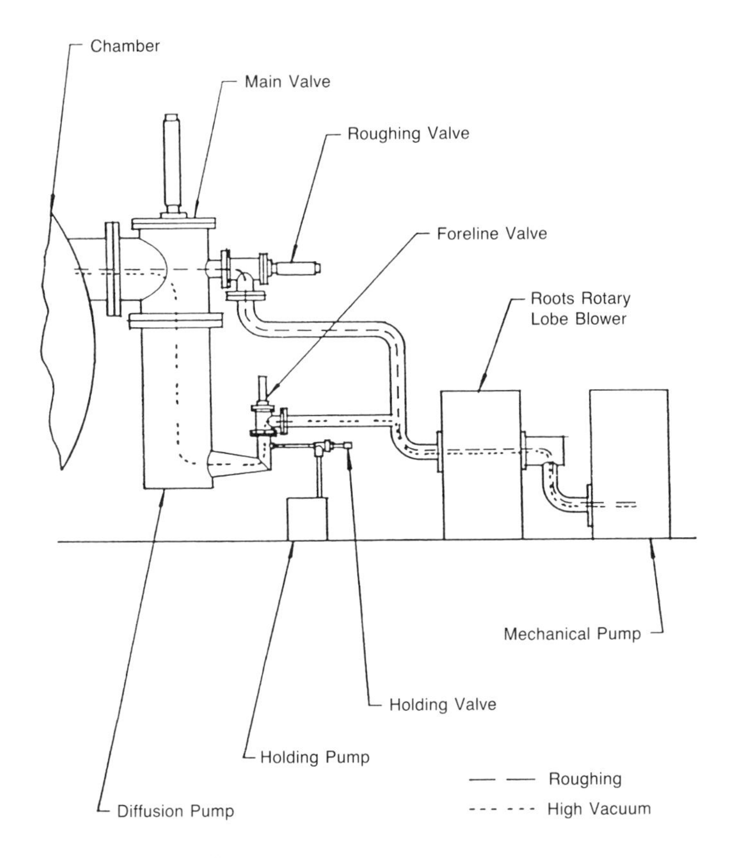

Fig 4.10 Vacuum pumping system

ous furnace. In general, where there is a high volume of similar workpieces, long production runs, and few different cycles, larger and more continuous furnace systems are favored. This may mean large batch units with automation controls linking them together in an integrated system, or continuous furnaces with dedicated material handling. As parts production moves more to just-in-time delivery or single-piece lots, smaller furnaces are favored. For around-the-clock production, automation can reduce the need for labor attention.

The second set of selection considerations to be addressed involves the design/performance factors related to the product being processed. These include part properties, such as strength and corrosion resistance; part geometry, such as size or precision; joining concerns; or the reactivity of the

Table 4.1 Furnace selection based on production considerations

Volume (Unit)	Low				High			
Length of run	Short	Short	Long	Long	Short	Short	Long	Long
Process	Few	Many	Few	Many	Few	Many	Few	Many
Batch furnace	Yes	Yes	No	Yes	Yes	Yes	No	Possible
Small	Yes	Yes	Possible	Yes	No	Yes	...	No
Large	No	No	Possible	No	Yes	No	...	Yes
Linked	...	Possible	...	Yes	...	Yes	...	Yes
Automatic	No	Possible	Yes	Yes	Yes	Yes	...	Yes
Dedicated*	Possible	No	Yes	No	Yes	No	...	No
Continuous	No	No	Yes	Possible	Possible	No	Yes	Possible
Small	...	...	Yes	Yes	Yes	...	No	No
Large	...	...	No	No	No	...	Yes	Yes
Linked	...	...	...	Yes	...	...	...	Yes
Automatic	...	...	Yes	Yes	Yes	...	Yes	Yes
Dedicated	...	...	Yes	No	Yes	...	Yes	No

*Cells or specialized controls

material selected for the part. All of these may be altered to suit a given furnace, but generally they are selected first by the designer, after which the proper furnace criteria must be determined.

Table 4.2 shows how these design/performance factors influence the choice of the system. As in the case of the production factors, it can be seen that as the design factors become less forgiving the range of furnace choices narrows. Higher temperatures favor higher-purity atmospheres with higher-vacuum capability. This is also the case when the workpieces are reactive. When temperature control is critical, with a narrow acceptance window, small batches are favored.

The third influence on furnace selection is quality. Of course, it is possible to produce the highest quality in even the simplest furnace, but what is meant here is the ability to simply control all quality management factors. For example, many specifications require that workpiece temperature be measured throughout the process cycle and recorded for permanent reference. It is almost impossible to satisfy this requirement using a continuous furnace. This would involve tracking many discrete small workpiece batches through the furnace, arranging for thermocouples to travel through a long furnace, and solving the problem of indexing them between chambers with vacuum-sealing doors. A batch furnace is obviously the preferred choice. In fact, in this case the vertical bottom-load furnace has an advantage

Table 4.2 Hot zone lining and pumping system selection

	Factors Favoring		
	Graphite/Ceramic	Refractory Metal Radiation Shield	Diffusion Pump
High Temperature	Equal	Equal	N/A
High Cooling Rate	Acceptable	Better	N/A
Uniform Cooling	N/A	N/A	N/A
Material Reactivity			
*Getter (e.g. Titanium)	Caution	Best	Yes
*Eutectics	Caution	Caution	N/A
*Outgassing	Possible	Best	Yes
*Carburizing/Sintering	Best	Possible	No
Critical Time at Temperature	Small Chamber	Small	N/A
Narrow Processing Range	Same	Same	N/A
Uniform Heating	Same	Same	N/A

in that work is loaded on the hearth outside the furnace, where it is easily accessed for thermocouple placement.

As the workpiece specifications become tighter, the batch furnace is also favored, with small furnaces being preferred to large ones, because there is inherently less variation in the process in smaller units. This is especially true if time at the processing temperature is very short, for example when austenitizing the high-speed tool steels. Likewise, factors such as the tolerance limits of the process, the scrap rate, or the risk resulting from a process error will put limits on both the lot size and the furnace selection. For instance, parts with a very high value before heat treatment will frequently be processed one at a time, to minimize the potential loss in the event of an error, even though it might be more economical to heat treat them in large batches.

One quality issue that is becoming more important is part cleanliness. This requirement affects not only the type of furnace selected, but also the operating practice. More chemically active materials, such as high-chromium steels, high-speed tool steels, stainless steels, or titanium alloys, require very-high-purity atmospheres. At low temperatures, e.g., in the range between 500 to 650 °C (930 to 1200 °F), it is very difficult to produce non-oxidizing conditions. During aging, stress relieving, or tempering, workpieces are often held for hours at these temperatures. After solution treatment or austenitizing prior to these treatments, the materials are chemically active and discoloration commonly occurs. For these reasons high vacuum is recommended and the preferred choice is a furnace with a lining that will not absorb contaminants or water vapor that could contaminate the vacuum chamber during the low-temperature treatment.

When the low-temperature cycle is a part of a higher-temperature cycle, for example a tool steel quench and temper, it is easy to program a batch furnace so that it is not opened until the end of the temper. While this ensures that it will not be contaminated by room air, it does not allow for a quality check between quench and temper. When such a check is needed, a continuous furnace may be the preferred choice, because having a vacuum chamber before and after the heating chamber, maintains the hot zone at the highest possible purity level, ensuring the best possible furnace atmosphere.

Auxiliary Equipment

As implied earlier, modern heat-treat systems are much more than simple batch furnaces with manual controls. Figure 4.11 shows an automated line with a combination of atmosphere and vacuum units. Weighing cost and quality production factors, this system was utilized by a manufacturer of high-volume small parts with many part numbers and a number of processes to be run randomly in a single system. To reduce labor content, the system was automated to allow for unattended running.

Vacuum was selected to provide quick cycle changes under short-run, short-cycle conditions, and for environmental reasons: The high-temperature process made it desirable to use a cold-wall furnace to minimize building heat load. Also, elimination of a combustible protective atmosphere also eliminated a source of heat and pollution. On the other hand, the low temperature of the tempering system, 190 °C (375 °F), made it possible and more economical to use conventional tempering furnaces. A compromise had to be made in the case of the workpiece cleaning equipment. The preferred technical choice would have been a solvent degreaser, but environmental concerns made this difficult. By using the proper detergent cleaner and hot drying cycle in the washer, it was possible to use a conventional water-based washing method which would adequately clean the workpieces before and after the high-temperature processing. This also permitted the use of a single washer rather than two in the line. With computer control of the work scheduling, there were no problems with mixing the processed and unprocessed loads.

It is generally difficult to justify vacuum furnaces for low-temperature processes such as tempering. As mentioned above, the low temperatures require high atmosphere purity. Also, radiation heat transfer is very inefficient at temperatures below red heat. To improve heating time, low-temperature furnaces often utilize the vacuum only as a purge prior to introducing an inert gas or reducing gas for convective heating. This further complicates the

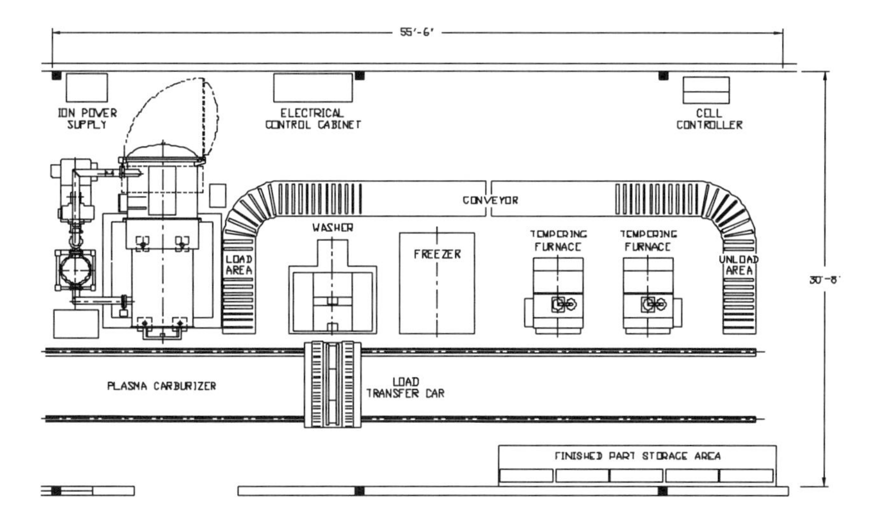

Fig 4.11 Automated vacuum furnace line with glow discharge

design, resulting in high relative cost compared with low-temperature atmosphere ovens.

Other Selection Considerations

Some of the reasons for selecting vacuum furnaces have been discussed above. The decision to purchase a vacuum furnace is sometimes unique to a given facility. It involves an evaluation of design, production, and quality factors. Compared with atmosphere furnaces, vacuum furnaces offer:

- Easily achieved neutral atmosphere capable of outgassing and oxide-free treatment of a wide variety of materials
- Higher temperature capability
- Rapid high-temperature range radiation heat transfer
- Cold-wall operation with minimal heat and exhaust input to the plant
- Quick changeover from one process cycle to another
- Quick start-up from a cold condition

The choice between batch and continuous furnaces is dictated by such factors as the percentage of time the furnace will operate per day, the volume of parts being processed, the variety of processes to be run, and quality issues such as the need to document temperatures of each lot. Ultimately, the cost of these choices determines the best-fit approach to a production situation.

While choice of the hot zone lining material has cost implications, more often it is dictated by technical concerns. The lowest cost lining is alloy elements and ceramic fiber insulation. This type of lining has limited temperature capability, is prone to distortion caused by expansion of alloys, tends to contaminate more rapidly, and is less thermally efficient in cooling. On the other hand, it is damage-tolerant, not easily destroyed by small leaks, and relatively easy to repair. As a result, it is frequently used in low-temperature processes such as annealing and tempering of steels.

Graphite-based hot zones are intermediate in cost, especially when the lining is a combination of a graphite hot face backed by a lower-cost ceramic fiber. These linings can operate to the 2100 °C (3810 °F) range, they exhibit low distortion because of the very low thermal coefficient of expansion of graphite, and they may not require the use of a diffusion pump because graphite acts as a "getter," or degassing agent. A good choice of graphite forms and shapes is available, which make these lining systems reasonably easy to maintain and repair. However, some materials are sensitive to carbon, and the lining material may trap evaporated gases or metals and contaminate subsequent loads. These linings have been very popular for sintering metals and nonmetals, carburizing, chemical vapor deposition, and oil-quench furnaces.

For the highest-purity processes at very high temperatures, the lining material of choice is refractory metal in a radiation shield type design. While the highest in cost, this type of lining is easy to pump down and outgas, it is carbon-free, it has a very low vapor pressure, and it can be evacuated to very high vacuum levels. In addition to their high cost however, these metals become brittle after exposure to temperatures much above 1100 °C (2010 °F), they require diffusion pumping or extensive inert-gas purging because they oxidize readily over 350 °C (660 °F) and they may form low temperature eutectics with the work and fixtures. These linings have been used extensively for furnaces for annealing and brazing superalloys, diffusion bonding, metallizing, and processing reactive metals such as titanium.

In choosing a lining material, no choice is universal. As a result, hot zones are often customized to individual user preferences. Combinations such as graphite heating elements and fiber linings, graphite elements and refractory metal shields, refractory metal elements and graphite linings, etc., are numerous. When based on knowledge of the properties of the lining material and vacuum design principles, such combinations can be successful, and a particular lining can sometimes be substituted for another.

Furnace Maintenance

Successful use of a vacuum furnace depends ultimately on the purity and reliability of its vacuum. Some basic maintenance procedures must be carried out to ensure furnace vacuum integrity.

Leaks

First, the furnace must be leak tight. Furnace leakage is easy to measure. After heating to remove entrapped contaminants, the furnace is isolated by closing the valves from the pumping system and measuring the time it takes for the furnace pressure to rise from one vacuum level to another. Leak-up rate, usually measured in microns per hour, should be checked routinely at least weekly. Tracking the leak-up rate over a period of time makes it possible to see any deteriorating trend. Even if the furnace is not completely baked, this test is valuable because it alerts the operator to a condition which may damage the furnace interior.

Leaks are the most time-consuming and troublesome of the maintenance items. Small leaks can only be isolated using helium type leak detectors. However, by monitoring where leaks most frequently occur, this factor can be minimized. Leaks can occur anywhere there is a penetration through the vessel wall, or at any major opening. The furnace door is sealed by an elastomer seal which is compressed against the mating metal faces by the pressure between the inside and outside of the vessel. This seal may become dirty from repeated opening, and it can harden and crack over time if exposed to excessive heat. It should be inspected and at least wiped clean regularly, or given a very thin coating of vacuum-grade grease. Similar treatment should be afforded other seals which may open and close regularly, such as the valve seals between the pumps and furnace chamber and any internal door seals in multiple chamber furnaces. These should be periodic maintenance items included in a quarterly, semiannual, and annual preventive maintenance program. Fittings and small valves normally need no maintenance other than a leak check if they are changed. Vacuum gages and thermocouples are consumable items. Because they are changed frequently, their feedthroughs tend to be major sources of leaks.

Check List Items

A major maintenance error is to attribute poor vacuum to leaks, when in fact the cause is either a defective gage or inadequate pump maintenance.

Mechanical roughing pumps and diffusion pumps are oil-filled, and their oil level is often the last thing an operator thinks of checking. To ensure good vacuum pumping, the pump oil must be of adequate quantity and cleanliness. One of the simplest preventive maintenance checks, which can provide a good warning of pump system oil level problems, is to record the furnace evacuation time on a regular periodic basis.

As with all furnaces, regular periodic lubrication of moving parts, calibration of instruments, and tuning of electronic instruments is necessary for quality operation.

Interior Lining and Heating Elements

The interior lining of the furnace should be inspected to eliminate hot or cold spots due to lining deterioration, and heating elements should be checked for loose connections. To minimize exposure of the furnace lining to humid room air, the furnace door should never be left open for extended periods. Also, keeping the furnace slightly warm between cycles prevents condensation in the lining, as does maintaining the water jacket at a temperature above the dew point of the surrounding room. Backfilling the furnace with dry, inert gas rather than simply venting it with room air also minimizes moisture entrapment. Likewise, the furnace should be evacuated when it will not be used for an extended period.

Cooling System Fluid

One item frequently overlooked is maintenance of the cooling system fluid quality. The cold-wall furnace, usually made of mild steel, must be protected from water-side corrosion. Corrosion inhibitors should be added to prevent corrosion and prevent buildup of deposits which can plug the cooling passages in the vessel shell and the heat exchangers. Synthetic coolants are often used to minimize this problem, but they must be checked for material compatibility.

Trays and Fixtures

Regardless of whether the furnace is a batch or continuous type, the workpieces are normally loaded on some form of support grid or tray. For low-temperature work these can be the conventional stainless steels or heat-resisting alloys used in atmosphere furnaces. For higher temperature work the most common materials used for work support are refractory metals, ceramics, or graphite. The major difference between atmosphere and vacuum furnaces is that in the extreme cleanliness of the vacuum environment these materials may readily react with the workpiece material. Sticking of fixtures to the work by diffusion bonding, and melting, due to the formation of eutectics are common vacuum furnace problems. One reason for this is that the processes they are used for, such as solution annealing, sintering, and brazing, take place very near the melting point of the materials being processed. The presence of any contaminant can lower the melting point into the treatment temperature range. Fortunately, such problems are generally restricted to temperatures over 1060 °C (1950 °F) and even then they are easily solved by selecting a proper fixture material or by using barriers such as ceramic pastes or cloth.

Special Atmospheres and Applications

Although the vacuum furnace is primarily a neutral-atmosphere processing furnace, reducing and reactive gases are frequently employed in it. Hydrogen is often used at a partial pressure. If it is provided with adequate safety devices to purge it and the pumping system and properly control the hydrogen, the same furnace can be used with either vacuum or hydrogen. The National Fire Protection Association (NFPA) has established strict guidelines for the use of hydrogen in vacuum furnaces. These are outlined in Standard NFPA 86D, Industrial Furnaces Using Vacuum as an Atmosphere, which is used by all vacuum furnace manufacturers. This standard can be obtained from the NFPA at P.O. Box 9101, Quincy, MA 02269-9101, or from any vacuum furnace manufacturer. In no case should hydrogen be used in a vacuum furnace without first reviewing the requirements of this safety standard.

Vacuum furnaces are frequently used for carburizing, nitriding, or coating. Those that are must be considered in a special class, because they must be equipped with special gas management systems. Furnaces used for vacuum carburizing employ a system which adds carburizing gas at reduced pressure with some means of promoting gas circulation, in one or more vacuum steps. Glow discharge or plasma carburizing furnaces operate on a similar boost and diffuse principle, but use special electrical isolation devices to cause the load to be negatively charged relative to the furnace interior. Using this principle, lower pressures with no requirement for circulation are possible compared with the vacuum carburizing approach. In addition to the isolation devices, this design requires a separate plasma generator power supply. Both types of furnace are most commonly constructed with graphite linings and heating elements and liquid-quenching equipment for hardening the carburized workpieces. Other than the fact that furnace design and the maintenance schedule must allow for soot buildup from the carburizing gas, these furnaces are identical to neutral processing vacuum furnaces. They must, of course, be equipped with the same safety aids required for hydrogen partial-pressure operation.

The above also applies to vacuum furnaces used for the other reactive-gas processing. Consideration must be given possible reactions between the gas and the furnace materials, safety requirements, and operating temperatures of the processes. Otherwise, selection, operation, and maintenance criteria are similar to those for other vacuum processes.

• 5 •

Work Preparation and Handling

Ross Pritchard, Vac-Aero International, Inc.

In the preparation and handling of workpieces for vacuum heat treating and brazing, there are several factors to be considered which can influence the end result of the processing. Cleanliness of the workpieces is important. They must be free of dirt, oil, grease, machining coolants, and forming compounds before they are loaded into the furnace. Some lubricants and cutting oils containing sulfur compounds can adversely affect the materials being treated. Inadequate cleaning can produce stained and discolored surfaces after heat treating or unsatisfactory braze alloy flow. If not removed, contaminants with high vapor pressures, such as drawing and stamping lubricants, will vaporize during heating, causing unacceptable increases in pressure and loss of vacuum. The vapors will condense on the colder surfaces of the furnace, presenting a potential for gassing off during subsequent higher temperature cycles.

Another consideration is the possibility of a reaction between the workpieces and the furnace fixtures, baskets, and trays. Depending on the chemical composition of the metals being treated and the heat treating basket materials, sintering of the workpieces to themselves and the baskets can occur. In the worst case scenario, a low-melting-point eutectoid alloy can be formed and severe damage to the workpieces and furnace can occur.

Finally some thought should be given to the loading of workpieces. Since heating in a vacuum furnace depends mostly on transfer of energy through radiation, workpieces should be spaced in the baskets so that the heating elements can radiate energy directly onto them. The shape and size of the part and its elevated temperature strength are also important considerations. Complex shaped components of alloys with relatively low strength at processing temperature may distort during treatment. In some cases, it may be necessary to support the part with fixtures made from high-strength, elevated-temperature alloys. A typical example is the annealing of thin sheet-metal workpieces made from iron-nickel high-permeability alloys such as Mumetal, Hypernik and Hy-Mu 80. At the annealing temperatures of 1120 to 1175 °C (2050 to 2150 °F), the yield strength of these alloys is quite low.

Cleaning

In most cases, workpieces are usually free of oxides, relatively clean, and ready for furnace loading when they arrive for vacuum processing. However, it is still good practice to check them for cleanliness before they are loaded into the furnace.

Machined parts, tubular assemblies, and castings with deep holes and recessed passages should be thoroughly inspected for trapped machining chips and other foreign matter. These contaminants must be cleaned out; otherwise passages may be blocked by sintering of machining chips, or other residues may react with part surfaces. Particular attention should be paid to castings. In some cases, foundries will clean castings in molten salt

Table 5.1 Cleaning methods for vacuum heat treating

		Suitability for Removing	
Method	**Bath Composition**	**Mineral Oil Cutting Fluids**	**Water-Soluble Oils**
Emulsion Cleaning	A mixture of insoluble hydrocarbons and water	Good	Good
Alkaline Cleaning	Water-base containing alkaline cleaners and other additives	Good	Good
Solvent Cleaning	- Mineral spirits - Alcohol - Acetone - Toluol - Chlorinated hydrocarbons	Good	Poor
Vapor Degreasing	Chlorinated hydrocarbons - Methylchloride - Perchlorethylene - Trichlorethylene - Trichloroethane	Excellent	Poor

Note: Do not use chlorinated hydrocarbons for cleaning titanium and zirconium alloys.

baths to remove core residues. If they are not thoroughly cleaned, entrapped salt can remain inside the castings. Since the salt vaporizes when heated in a vacuum, it must be removed before processing.

Another word of caution: metal identification tags are often wired to parts destined for vacuum heat treating. Both the tag and the attachment wire should be checked to ensure that they are not made of alloys such as aluminum, which may melt during the furnace cycle.

Table 5.1 lists methods commonly used for cleaning workpieces for vacuum processing. Selection of the cleaning process will depend to a large degree on prior processing history and the type of alloy to be treated. With machined, stamped, or pressed parts, it is important to know what type of cutting fluid or die lubricant was used in the manufacturing process. Some of these fluids are mineral-based oils, while others are water-soluble oils. Performance of the various cleaning processes depends on the nature of the contaminant.

Mineral-Oil Cutting Fluids

For workpieces which are lightly soiled with mineral oil cutting fluids and polishing compounds, solvent immersion cleaning or solvent vapor degreasing, using spray to remove machining chips, is usually satisfactory. Both alkaline and emulsion cleaning can also be used for these oils.

Water-Soluble Oils

Solvent or vapor degreasing may not remove water-soluble cutting fluids. Moreover, water contained in the fluids can cause breakdown of the solvents used in vapor degreasing, to form acids which can severely attack both the workpieces and the equipment. For workpieces contaminated with water- soluble oils, alkaline or emulsion cleaning effectively removes these substances.

When the contaminant composition is not known, solvent cleaning or vapor degreasing is the first choice for cleaning the workpieces. If they do not clean up, try using alkaline or emulsion cleaning as an alternative method.

Special Cleaning Precautions

The following are special precautions to be followed when cleaning workpieces for vacuum heat treating:

Titanium and Zirconium. Titanium, zirconium, and their alloys should not be cleaned in trichlorethylene, trichloroethane, perchlorethylene, or methylchloride. All of these solvents contain chloride compounds, and chloride residues can cause stress-corrosion cracking in titanium and zirconium alloys when they are heated above 280 °C (550 °F). Chloride-free solvents such as acetone, toluol or alcohol, or alkaline cleaning should be used instead. After cleaning, the workpieces should be handled with clean, white gloves to prevent fingerprints. Other nonferrous and ferrous alloy workpieces can be safely cleaned in chlorinated solvents.

Nickel Alloys. Lubricants used for deep drawing may contain lead, sulphur, and other low-melting elements, and these must be removed. These elements, particularly sulphur, can attack nickel alloy surfaces during heating, eventually forming low-melting eutectics which can severely embrittle the base metal.

Drying After Cleaning

Regardless of the cleaning method used, workpieces must be dry and free of cleaning solution before they are loaded into the vacuum furnace. Solvent and water cleaning residues will volatilize during heating, eventually collecting in the vacuum pumps. Contamination degrades pump performance, and furnace pump down times can increase significantly.

Safety Considerations

Certain safety, health, fire, and environmental factors must be considered with all cleaning processes. Regulations specified in applicable Occupational Safety and Health Agency (OSHA), Environmental Protection Agency (EPA), and National Fire Prevention Agency (NFPA) standards must be met when operating and using any cleaning process. These standards are constantly being reviewed, and processes that are currently acceptable may be banned in the future. For instance, production of fluorocarbon 113, widely used for ultrasonic degreasing, and 111 trichloroethane, a commonly used vapor degreasing solvent, are soon to be phased out because they tend to deplete the atmospheric ozone layer. From an environmental standpoint, trichlorethylene and perchlorethylene will still be produced for use in vapor degreasers, but the permissible OSHA exposure limits for both these solvents are much lower than for trichloroethane. Both good ventilation and worker protection must be provided when operating equipment with these solvents.

Fixturing

Fixture materials and design must be appropriate for the application, and fixture maintenance procedures are also important to successful vacuum processing.

Fixture Materials

Materials suitable for vacuum furnace fixtures and baskets are listed in Table 5.2. Selection of fixture material is determined by cost, elevated-temperature properties, the type of quenching required (inert-gas or oil), and compatibility with the workpieces and furnace hearth materials. Since the fixtures are used in a vacuum environment, resistance to oxidation and other high-temperature chemical attack are not important considerations.

Gas-Quenching

For service temperatures to 980 °C (1800 °F), stainless steels such as Types 304, 309 and 310 are often used for baskets. These alloys can become embrittled from carbide precipitation and sigma formation during long exposure at temperatures between 595 and 815 °C (1100 and 1500 °F). More stable alloys such as 35Ni-15Cr or Inconel 600 are usually selected for service in this temperature range. Although more expensive than the 300 Series alloys, the longer life of these superior alloys may offset the extra cost.

Alloys with higher elevated-temperature properties are required for heavy loading and long life at temperatures above 1150 °C (2100 °F). Haynes 230 alloy provides reasonable life to 1200 °C (2200 °F); however, for service above this temperature, nickel-base oxide-dispersed alloys such as MAP 956

Table 5.2 Fixture materials for vacuum heat treating

Material	Elevated(1) Temperature Strength	Thermal Shock(1) Resistance	Relative(1) Cost
304 SS	1	1	1
309 SS	2	3	2
310 SS	2	3	2
35Ni-15Cr(2)	3	6	4
Inconel 600	2	5	5
Inconel 617	4	6	6
Alloy 230(3)	5	7	7
MA 956(4)	7	8	8
Molybdenum	10	8	10
Graphite	9(5)	8	3

Note: (1) Rated on a scale of 1 to 10 with 10 being the highest or most desirable rating. (2) 35Ni-15Cr alloy includes a range of alloys containing nickel from 30% to 40% and chromium from 15% to 23%. Included in this group are RA330, Incoloy 800 and other proprietary alloys. (3) Haynes 230 nickel-base alloy. (4) Proprietary oxide-dispersed nickel-base alloy made by International Nickel. (5) Graphite should be used in compressive loading. Compressive strength of graphite at 1260 °C (2300 °F) is approximately 9000 psi.

or pure molybdenum are recommended. Molybdenum has excellent high-temperature properties. The longer life and energy savings resulting from the use of lighter molybdenum fixtures may justify their much higher cost. Molybdenum recrystallizes when exposed to temperatures in excess of 1150 °C (2100 °F) and is brittle at room temperature. For this reason, molybdenum fixtures must be treated with care during loading and unloading.

For some specialized applications, graphite fixtures are used. The strength of graphite increases with temperature; it has good thermal shock resistance; it has reasonably high thermal conductivity; and it is relatively inexpensive. These unique properties make graphite the ideal fixture material for supporting workpieces with large, flat surfaces, where flatness must be maintained during heating. To prevent possible carbon pick-up in the materials being treated, graphite surfaces in contact with the workpieces should be coated with a suitable stop-off material.

Oil-Quenching

Oil-quenching baskets and fixtures should be constructed of materials with good thermal shock properties and high resistance to carbon pick-up. Heat-resistant alloys such as the 35Ni-15Cr series are commonly used for oil-quenching operations. If cracking due to thermal shock is a problem, the higher-nickel alloys such as Inconel 617 and Haynes 230 can be used.

Fixture Design

The goal in fixture design should be minimum weight and good service life at minimum cost. Reduced weight is important for both economic and processing performance reasons. The lower the fixture mass, the less energy required for heating, resulting in faster heating of the gross load. During gas quenching, lower mass means there is less energy to be extracted from the load, and faster cooling rates can be achieved.

The stainless steels and most of the nickel alloys are readily weldable; however, oxide-dispersed nickel alloy MA 956 and molybdenum are not weldable, and fixtures made from these materials must be designed for mechanical fastening rather than welding.

When designing heat treating or brazing fixtures which clamp or restrain the part, careful consideration must be given to differences in thermal expansion between the part material and the fixture material. Wherever practical the coefficient of thermal expansion of the fixture material should closely match the coefficient of the alloy being treated. If this is not possible, the fixture must be designed to accommodate the dimensional differences induced by heating.

Fixture Maintenance

It is good practice to vacuum-bake fixtures at a temperature at least 25 °C (45 °F) higher than their maximum operating temperature, to degas them before they are used in production. If the fixture is of welded construction, all welding fluxes must be removed before baking. With repeated service, fixtures and baskets tend to become discolored. Periodically they should be cleaned by abrasive blasting or by subjecting them to a hydrogen-partial-pressure clean-up cycle at 1150 °C (2100 °F).

Workpiece Loading

Workpieces should be loaded into clean, degassed fixtures. Heating of the workpieces in vacuum occurs as a result of energy transferred by radiation from the heating elements of the furnace to the workload, and the workpieces should be spaced to expose as much as possible of each part to direct radiation from the elements. In densely packed loads, workpieces tend to shield one another, retarding heating rates. It is good practice to attach load monitor thermocouples to the center and outer surfaces of the load, to determine when it reaches soak temperature. If it is not possible to use load thermocouples, longer soak times must be used on dense loads, particularly at temperatures below 760 °C (1400 °F), where heat transfer by radiation is significantly inefficient.

Adequate spacing is also important if the workpieces require gas-quenching. Without sufficient clearance between them, quench-gas flow can be restricted, resulting in inadequate cooling and failure to meet required properties.

Eutectic Reactions

In a vacuum furnace environment, metal surfaces remain clean and free of oxides. These active surfaces are conducive to diffusion of certain elemental combinations with a high tendency for interaction. Table 5.3 lists metal combinations which readily diffuse and form lower-melting-point eutectics if these materials come into contact with each other. Titanium-nickel and graphite-nickel are two combinations which should be avoided. Nickel and titanium form a eutectic at 955 °C (1750 °F). Although most titanium heat treatment temperatures are lower than 955 °C (1750 °F), care should be taken to ensure that all titanium workpieces and machining chips are removed from stainless steel and nickel alloy baskets after completion of the treatment, particularly if the baskets are used for processing other work at high temperatures. Similarly, carbon (graphite) will react with nickel to form a eutectic at 1125 °C (2060 °F). For this reason graphite used in furnace hearth

Table 5.3 Metal combinations producing eutectic reactions

Combination	Eutectic Formation Temperature °C (°F)
Nickel-Titanium	943 (1730)
Nickel-Zirconium	961 (1762)
Nickel-Carbon(1)	1316 (2400)
Nickel-Tantalum	1320 (2410)
Molybdenum-Nickel	1320 (2410)

Note: (1) Carbon (graphite) reacts with nickel-base alloys and nickel-containing stainless steels to form eutectics as low as 1125 °C (2060 °F).

rails and fixtures must not come into direct contact with nickel-bearing baskets or workpieces being heated above the eutectic temperature.

High purity, stable ceramic materials such as aluminum oxide (alumina) and zirconium oxide (zirconia) can be used to separate reactive metal combinations during vacuum processing. These materials are used in several different forms, including paints, thermally-sprayed coatings, papers, cloths, powders, and sintered shapes.

Paints

Stop-off paints consisting of ceramic powder, usually aluminum oxide (Al_2O_3) or zirconium dioxide (ZrO_2) of 99.4% minimum purity, and a suitable acrylic plus a solvent vehicle, are often applied to fixtures to prevent sintering and braze alloy wetting. For work where sintering is a severe problem, the painted fixtures should be subjected to a vacuum heat-treat cycle at a temperature higher than that specified for the production cycle, to bake in the paint before being used for actual production. Fixtures should be inspected after each run, and areas where the stop-off is worn should be touched up before re-use.

Plasma Coatings

Alumina and zirconia plasma-sprayed coatings are thicker and more durable than stop-off paints. Typical examples of commercially available plasma coatings which provide excellent protection and reasonable life are Metco 204 (zirconia) and Metco 101 and 105 (alumina). Other plasma powder manufacturers can supply equivalent coatings. The zirconia coatings are three to four times more expensive than alumina, but zirconia coatings tend to have better thermal shock resistance.

Ceramic Paper, Cloth, and Powder

High alumina-silica papers and cloths are also used to separate workpieces and fixtures. Different grades of cloth and paper are available. Some contain volatile binders used in their manufacture. Binder-free types are preferred. If binder-containing materials are employed, they must be vacuum-degassed before being used in production. Ceramic papers should be used sparingly because they tend to break down during gas-quenching. The loose, abrasive fibers are blown around during quenching and can eventually be sucked into and seriously damage mechanical vacuum pumps.

High-purity alumina powders are often used to separate high-nickel magnetic shielding alloys during vacuum annealing. The workpieces are embedded in fine alumina powder (–250 mesh) contained in stainless sheet metal containers. The containers should have suitable covers to prevent dispersal of the powder during quenching.

Sintered Ceramics

Sintered, high-purity ceramics in the form of plates and bars are excellent materials for separating problem material combinations. Round ceramic bars are placed in grooves on the upper surface of graphite hearth rails to prevent heat-treating fixtures from contacting the graphite. Dense, fine-grained ceramics of high-purity alumina or yttria-stabilized zirconia are two of the more commonly used materials. The more expensive zirconia has a higher maximum temperature limit, 2200 °C (3990 °F), than alumina, at 1950 °C (3540 °F), although for most normal vacuum heat-treating processes alumina is satisfactory.

Summary

The cleaning, preparation, and loading of workpieces is a critical operation in the vacuum heat-treating process. Inadequate cleaning and improper loading and fixturing can adversely affect the outcome of the heat-treating process, and in extreme cases, such as eutectic melting, can seriously damage the furnace equipment. Operators should receive thorough instructions regarding correct cleaning and loading procedures and should be trained to recognize potential problems before they occur. Care and caution taken during this operation can prevent problems from occurring later in the process.

• 6 •

Principle Process Variables

William R. Jones, Vacuum Furnace Systems

In heat-treating or brazing metals, measurement and control of temperature is a foremost consideration. In order for the desired reaction or metallurgical conversion to be properly carried out, e.g., to correctly austenitize a tool steel or to fully melt a filler metal and cause it to flow and properly braze a joint, time at temperature is critical. Exposure to heat and to the vacuum furnace residual atmosphere or a controlled partial-pressure gas such as hydrogen, can either cause a specific desired reaction, such as oxide reduction and workpiece clean-up, or it can damage the surface of the work. The method of cooling, as in a programmed cool to achieve a full anneal, or a full-blown 10-atm positive-pressure quench to achieve specific metallurgical effects, becomes an all-important consideration once a part is at temperature and properly soaked out. This chapter addresses these subjects.

Temperature Measurement

The operator of a vacuum furnace must be concerned with two important temperature measurements: the temperature of the furnace hot zone, used to control heat input, and the temperature of the actual workload undergoing heat treatment.

Furnace Heating

In the modern vacuum furnace, hot-zone temperature measurement is normally accomplished with thermocouples located in the vicinity of the heating elements, per Fig 6.1. Normally a minimum of two are employed, one for furnace control and a second, connected to an independent hot-zone power supply shutdown and alarm, for over-temperature control. In practice, the size of the hot zone may dictate multiple zone control, with multiple

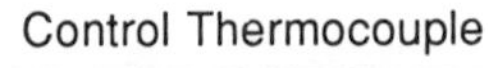

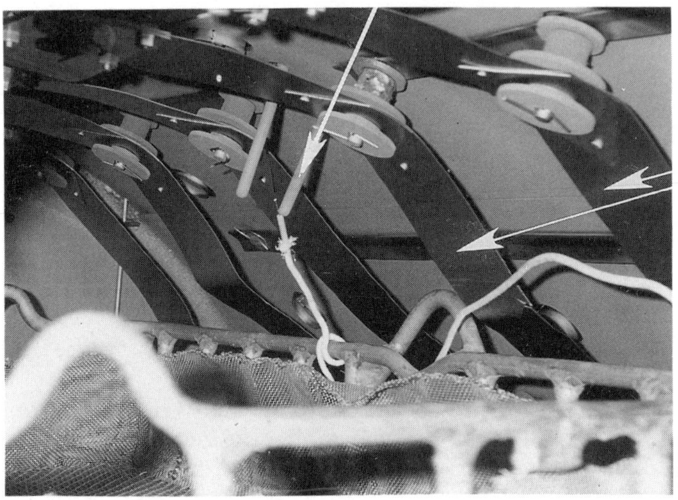

Fig 6.1 Heating element and control thermocouple

Table 6.1 Temperature range of standard thermocouples

Type	Alloy	Range °C (°F)
J	Iron	Ambient to 760 °C (1400 °F)
K	Nickel	Ambient to 1260 °C (2300 °F)
N	Nickel	Ambient to 1260 °C (2300 °F)
S	Platinum	65 °C (150 °F) to 1480 °C (2700 °F)
R	Platinum	65 °C (150 °F) to 1480 °C (2700 °F)
B	Platinum	205 °C (400 °F) to 1705 °C (3100 °F)
W	Tungsten	205 °C (400 °F) to 2315 °C (4200 °F)

For further reference see ASTM E230 and E988.

thermocouples for each function. Where process temperatures exceed normal thermocouple ranges, per Table 6.1, or in special situations such as non-contacting applications in a semi-continuous vacuum furnace, an optical pyrometer may be employed. Often optical pyrometer temperature readings require correction because of furnace sight-glass errors, which come about because of plating on the glass, from workpiece evaporation. Also, objects under observation may deviate from black-body emissivity conditions, resulting in severe absolute-temperature measurement errors.

Hot-zone control thermocouple signals, in the millivolt range, are generally transmitted to a process temperature controller-programmer, often connected in parallel with temperature recorders, data logging instruments, or computers. The over-temperature thermocouple signal may be tracked accordingly, as a back-up record of hot-zone temperature.

Thermocouples. For normal processing temperatures in the 205 to 1455 °C (400 to 2650 °F) range, a widely accepted control thermocouple is the platinum Type S, and in some cases platinum Type R, housed inside a ceramic (alumina) or molybdenum tube. For higher temperature ranges up to 2205 °C (4000 °F), tungsten Type W thermocouples may be used. For operating temperature ranges of 980 °C (1800 °F) and below, nickel Type K or N thermocouples are used. All control and over-temperature thermocouples should be of specified premium-grade wire, calibrated to one or more points, depending on the application or process standard. Traceability to the National Institute of Standards and Technology (NIST) is required for aircraft certification, military specifications, or nuclear applications.

Thermocouple Locations. Location of the control and over-temperature thermocouples within the furnace hot zone requires attention to several factors. The first for most applications is to insert the thermocouple 50 mm (2 in.) past the heating elements into the effective hot-zone envelope. In low-temperature applications, such as aluminum brazing and tempering up to 650 °C (1200 °F), thermocouples are often connected directly to the heating elements to achieve rapid thermal response. Otherwise thermal response lags as a result of low thermal radiant energy at these temperatures.

Workpiece Temperature Tracking

Until recently actual workpiece temperature measurement with thermocouples was difficult, due to cumbersome wire feed-through glands and failure-prone internal thermocouple plugs and jacks. Relying instead on the heat treater's rule of thumb: "one hour per inch of thickness," a furnace operator brought the furnace up to temperature and soaked out time at temperature accordingly. While this old rule is surprisingly accurate in the higher heat ranges, i.e., 980 °C (1800 °F) and above, at lower temperatures, particularly in vacuum, it can cause serious underheating. Consequently, operators tend to over-compensate, and this has resulted in much wasted furnace time. Use of workload thermocouples eliminates the guesswork and brings the heat-treating process under control, resulting in a higher quality product and often in reduced processing time.

Problems with thermocouple wire feed-through glands were solved in 1980 (see Fig 6.2), and satisfactory internal thermocouple plugs and jacks

Fig 6.2 Vacuum feed-through assembly

have been available on vacuum furnaces manufactured since. Older furnaces can be retrofitted to upgrade them with these devices.

The use of these temperature tracking workpiece thermocouples, correctly placed in the workload makes it possible to record exact time at temperature for each part that is processed, as shown by the sample reproduced in Fig 6.3. This makes it possible to base process decisions on actual temperature records, to measure possible overheating of outer workpieces

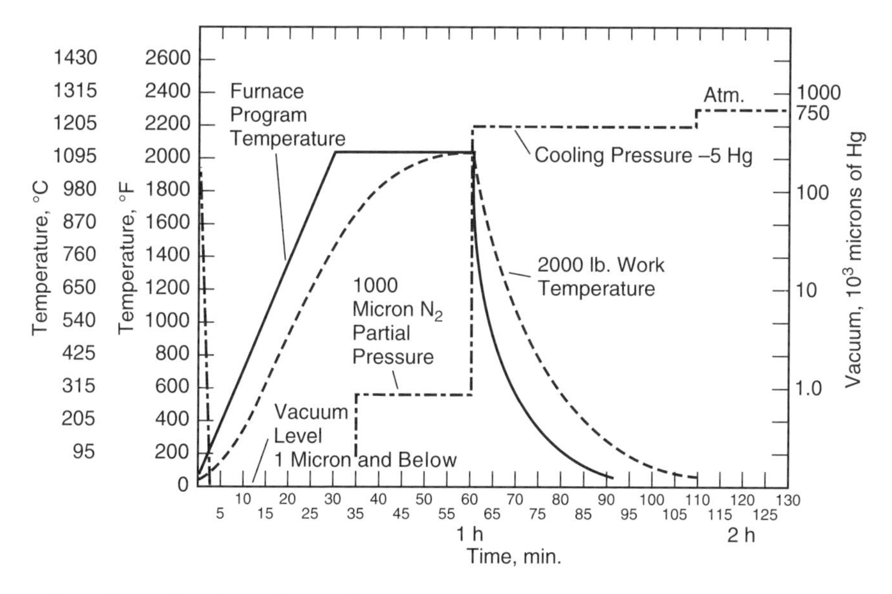

Fig 6.3 Typical N_2 brazed valve body run

and bring it under control, etc. It is important that all parts of the workload reach process temperature in as close to the same time as possible, and to accomplish this, several intermediate temperature soaks may be required before the final temperature is attained. This is particularly true when attempts are being made to minimize distortion or cracking in complex machined workpieces of deep-hardening steels, such as H11 and H13 tool steels. Brazing applications also may require intermediate soaks to effectively melt the braze alloy throughout the workload and to prevent excess flow, diffusion, or surface erosion of overheated parts at outer locations.

Thermocouples. Work thermocouples are generally made with Type K or N wire, with flexible high-temperature Refrasil insulation, mullite ($Al_6Si_2O_{13}$) spacers, or stainless steel sheaths over crushed magnesium oxide insulation. Depending on design, a temperature measurement range from ambient to a maximum of 1315 °C (2400 °F) is possible with these thermocouples. But when used extensively at temperatures over 1205 °C (2200 °F), they tend to develop a temperature shift, due to high-temperature aging, which limits their service life. Measurement of work temperatures over 1315 °C (2400 °F) requires the use of thermocouples made from platinum or tungsten wire, and the fragile nature of fully-annealed platinum wire or recrystallized tungsten wire must be allowed for. Optical pyrometers

(a)

(b)

Fig 6.4 Workload thermocouple placement

can also be employed, requiring sight-glass and observation emissivity corrections as discussed above.

Thermocouple Locations. The locations where work thermocouples are placed in the process load can seriously affect temperature indications. These thermocouples should be placed deep inside the workload, and where possible inserted into either existing holes in the workpieces or holes drilled especially for the purpose when this can be done. Drilled holes should be directed to the center of the part where practical. A second thermocouple can be placed midway between the center of the workload and the outer workpieces, and a third thermocouple at the surface of the workpiece or outermost part of the workload, per Fig 6.4(a) and (b). Other thermocouples may be placed as desired, with complex configurations utilizing as many as 12 to 24 of them. These work thermocouples transmit input signals, in the millivolt range, to process recorders, data loggers, or computers. As the furnace hot zone is subjected to soak temperatures, the outer workpieces and those with thin cross sections come up to temperature first, intermediate workpieces second, and finally those located deep in the load, along with thicker sections of the larger workpieces.

Vacuum and Atmosphere Control

The vacuum furnace chamber is completely sealed, and almost completely leak-free. Compared with a normal atmospheric furnace it is totally airtight, allowing high-vacuum operation, and, when gas atmospheres are injected, they can be controlled to a high degree of purity. However, these conditions exist only if the vacuum furnace is under a regular maintenance program, with a quality assurance program that maintains a log of vacuum pump-down performance, and regular, periodic leak-up rate tests (Ref 1).

Vacuum Processing

Work to be processed in a vacuum furnace must be free of oil, heavy oxides, etc., or contamination of the hot zone and the work could result. The work is loaded into the furnace, the work thermocouples are attached, and the furnace door is closed and locked or bolted. The vacuum pump valves are opened and the vacuum chamber is pumped down to operating pressure. Roughing pumps pump down to 10^{-1} torr, or with the addition of a Roots vacuum blower, 10^{-2} torr, or to 10^{-5} torr and below with the further addition of a suitably-sized high-vacuum oil diffusion pump.

Note that a pressure of 1×10^{-3} torr (one micron H_g) is approximately equivalent to gas purity of 1 ppm at atmospheric pressure and an equivalent dewpoint of –75 °C (–100 °F). Vacuum furnaces normally operate at high vacuum (below 1×10^{-3} torr), and this is the equivalent of a better purity level than can normally be delivered to the plant by commercially purchased bulk liquid argon or nitrogen gas (Ref 2). Pump-down time for the vacuum furnace is usually fast; 1×10^{-5} torr in 1 hr is common. Compare this with a sealed atmospheric retort furnace in which air is purged from the chamber with an inert gas: reaching 100% purity is impossible and to reach a ppm level comparable to the high-vacuum atmosphere referred to above would require 24 hours and a great deal of gas purging.

In the vacuum furnace, when the workload is heated, air, moisture, oils, solvents, and other gases are liberated. In addition the furnace hot zone outgasses as well (Ref 3), because it is exposed to the ambient air each time a new processing work cycle begins. The outgassing causes the pressure to rise, and to maintain a specific minimum vacuum level, e.g., 5×10^{-4} torr, heating rates must be controlled. Suitable soak times or holds must be programmed to maintain the work temperature and process vacuum and allow the vacuum pumps to recover from the resultant outgassing. A vacuum furnace should never be heated at full power directly to operating temperature without consideration of vacuum levels, possible distortion damage to the work, and excessive thermal shock to the hot-zone materials. Full-power heating also makes heavy electrical power demands and higher-than-necessary power costs. Input power limits can be programmed into the process controller to minimize this problem.

Because the vacuum pumps are large and constantly pumping on the vacuum chamber, hot zone, and work, undesirable released gases are quickly removed from the vacuum chamber. For a better appreciation of the high-purity operating performance of the vacuum furnace, contrast this with normal inert-gas atmosphere furnace practice, in which outgassing contaminants are partially purged by the inert gas atmosphere, which become diluted by the unpurged residue.

Partial-Pressure Processing

High-vacuum operation is not always desirable. Base metal elements such as nickel and chrome can evaporate from workloads when processed in high vacuum and at elevated temperature (see Fig 6.5). Temperatures above 1090 °C (2000 °F) and hold periods longer than 30 min. are conditions in which evaporation of base metals or alloying elements can occur. To prevent it, an inert gas such as argon or nitrogen can be injected into the

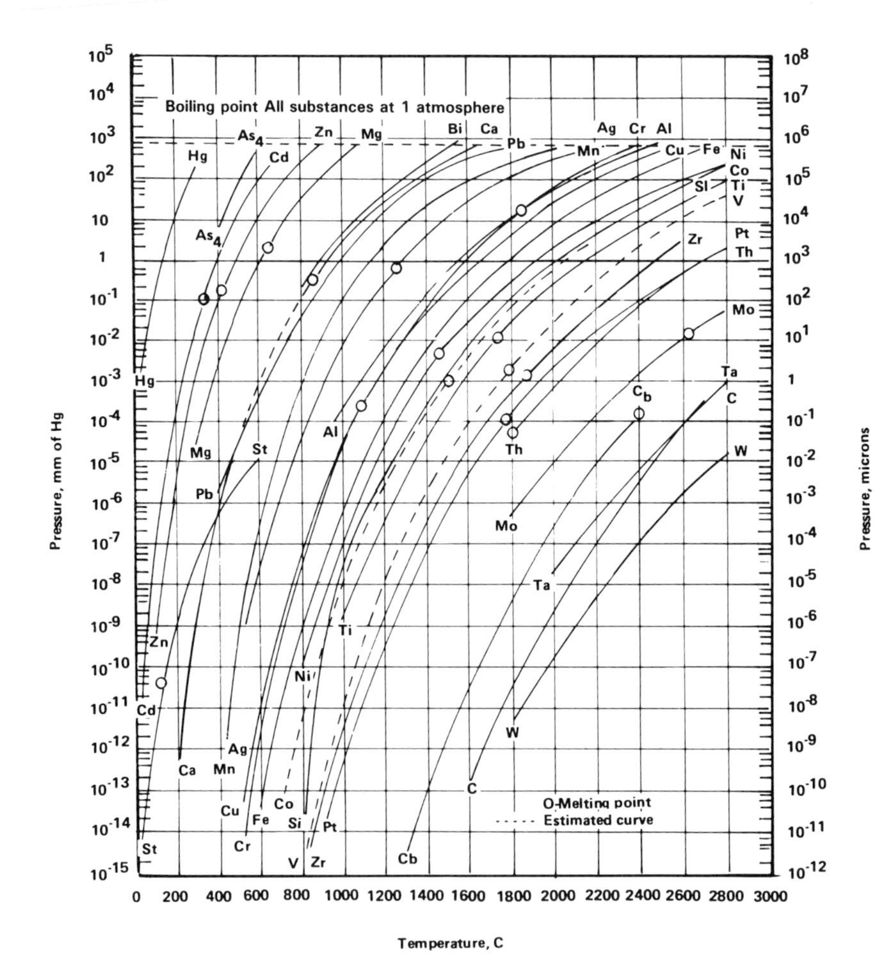

Fig 6.5 Vapor pressure of the elements.

vacuum furnace chamber through a controlled bleed valve, until the pressure has risen to between 10^{-1} and the 10 torr range.

Higher pressure, between 1 and 10 torr, is often desirable to positively reduce evaporation, and also provide a sweep gas flow through the furnace to improve hot-zone gas purity. Operation near blank-off (about 10^{-1} torr) with roughing pumps, and at elevated temperature above 760 °C (1400 °F), should be avoided because of the possibility of contaminating the hot zone materials of construction, or sensitive process work, such as stainless steel type 718. This contaminating effect of residual gases is a result of the water vapor and vacuum pump oils usually associated with mechanical vacuum pumps operating near blank-off.

Gases. Process gases such as argon, hydrogen, and nitrogen can be introduced into the vacuum furnace hot zone to carry out a specific purpose. In the case of hydrogen, the normally near-neutral vacuum atmosphere can be sharply shifted to a reducing atmosphere, to carry out a reducing reaction or to prevent oxidation of sensitive process work. Hydrogen should not be introduced until the furnace has been fully evacuated to below 10^{-1} torr, and the temperature is over 760 °C (1400 °F). This must be done in full compliance with the National Fire Prevention Agency (NFPA) 86D safety standard, because hydrogen is a well-known explosion hazard. Never operate the vacuum furnace over 15 torr hydrogen pressure or at temperatures below 760 °C (1400 °F). To do so is a violation of this standard and other safety regulations.

A further caution, this one of a totally different nature, concerns the use of nitrogen gas for partial-pressure cooling or gas-quenching. Certain superalloys or stainless steels, such as Grades 321 and 718, contain trace amounts of titanium or other elements that may react with standard atomic nitrogen (N_a) gas to cause surface nitriding. Yellowing of the work will result. In these cases completely inert argon gas may be substituted for nitrogen gas.

Brazing. In brazing steel or stainless steel materials with copper, the brazing temperature is normally 1120 °C (2050 °F), at a recommended nitrogen partial pressure of 5 to 10 torr. Workloads are heated in a full vacuum (below 10^{-1} torr) to 760 °C (1400 °F), and partial-pressure gas flow is introduced through the brazing preheat temperature and finally to the completion of the brazing temperature. Evaporation of copper and contamination of the hot zone from it is minimal with this procedure. (For other brazing alloys, see Chapter 2.)

Pressure Measurement. When operating the vacuum furnace with partial-pressure gases, the particular type of vacuum gage and the effect of the gas on its calibration must be considered. Thermal-type vacuum gages are calibrated for air or nitrogen gas. Because other gases will cause their calibration to shift radically, it is advisable to use an absolute pressure gage, such as an electronic capacitance manometer, for measurement when partial-pressure processing. Other alternatives are to use a thermal vacuum gage with built in calibration curves specific to the kind of gas and range of pressure the process requires (Ref 4), or to obtain specific calibration data over the desired vacuum pressure range from the vacuum gage manufacturer.

Cooling and Quenching Control

In cooling or gas-quenching, parts with thin sections or located near the edges of the hot zone will cool first, and those with heavier sections or at inner locations in the workload will cool last. Using work thermocouples makes it possible for the operator to decide, based on actual temperature indications, when to open the furnace to the air and remove the work. Sensitive materials such as tantalum, titanium, or powdered metals that may be pyrophoric must often be kept in the protected gas until they have cooled to temperatures near ambient, while some alloy steels or tool steels such as H11 or H13 can be air-released at 205 °C (400 °F) or higher for applications in which some slight oxidation, yellowing, or blueing, is permissible. Concern for the furnace hot zone and materials of construction is also a consideration. Vacuum furnaces with molybdenum or graphite radiation shields (see Chapter 4), are not normally opened when hot-zone components are above 205 °C (400 °F).

Integral Liquid-Quench Furnaces

Some vacuum furnaces are built with integral oil- or water-quench chambers. This type of furnace is used for processing alloy steels and metals that require extremely rapid quenching, with cooling rates above 1120 °C/min. (2000 °F/min.), such as water- or oil-hardening W or O grade tool steels and 4140 and similar alloy steels. In such furnaces, after the work is heated and thermal-soaked in vacuum, it is quickly transferred to the quench chamber, which is backfilled with inert gas, and submerged in oil or water, which is usually agitated to further increase the cooling speed. This cools the work quickly to near-ambient temperatures at which it can be removed from the furnace.

Gas-Quenching

There are many air- or oil-/air-hardening alloys that can be cooled or quenched satisfactorily in gas. Gas-quenching can achieve cooling rates of 280 °C/min. (500 °F/min.) and development efforts are underway to improve these rates further. Normally all the air-hardening grades of tool steel are processed in gas-quench vacuum furnaces, including the popular A2, D2, M1, S7 Grades and the hot-work H11 and H13 Grades. The 300 and 400 Series stainless steels are also gas quenched, as well as many of the stainless steel and titanium superalloys.

Gas quenching follows the final thermal soak in vacuum. The furnace is quickly backfilled with inert gas to atmospheric or positive pressure, and the gas, driven by a powerful fan or blower and continuously recirculated, flows at high velocity over the workload and through a gas-to-water heat exchanger (Ref 5). Gas-quench fans or blowers range from small 300 ft^3/min., 5 hp fans for smaller furnaces, to 50,000 ft^3/min., 250 hp blowers for larger, high-performance furnaces.

Gases. The inert gases used for gas quenching are nitrogen, argon, and helium. Nitrogen, the most popular, is the least expensive. Argon is completely inert and almost universally specified by the aircraft, nuclear, and medical industry. Because it has poor thermal heat transfer characteristics, argon cools 30% more slowly than nitrogen, and it can cost four to five times as much per cubic foot. Helium cools 10% faster than nitrogen and, like argon, is completely inert and an excellent cooling gas, but unfortunately it costs up to 25 times as much as nitrogen. Hydrogen, like helium, has excellent heat transfer characteristics, but since the furnace must be cooled to ambient and opened to air, the risk of premature opening through operator error forbids its use for this purpose, per the NFPA 86D safety standard.

Positive-Pressure Gas-Quenching

There is a growing trend to enhance quench rates through the use of cooling gas at greater than atmospheric pressures of 60 psig or 5 bar (Ref 6). Some designs contemplate operation to 1485 psig or 100 bar. The advantage of higher pressure cooling is a denser gas, with increased mass flow and therefore greater thermal conductivity, all of which add up to improved cooling rates (see Fig 6.6). In addition, gas blowers and heat exchangers operate at better efficiency at increased pressures.

Equipment Requirements. Vacuum furnace chambers may operate to 15 psig, 2 atm, or 2 bar and generally stay within the American Society of Mechanical Engineers (ASME) pressure vessel code without requiring a code stamp. Chambers operating over this pressure must be designed, manufactured, and tested to the specific ASME code pressure. Code-stamped vacuum chambers can cost anywhere from 2 to 10 times as much as a standard vacuum chamber, depending on the specific over-pressure requirement. However, improved performance, including reduced overall cycle time, can justify this added cost for some applications, e.g., processing

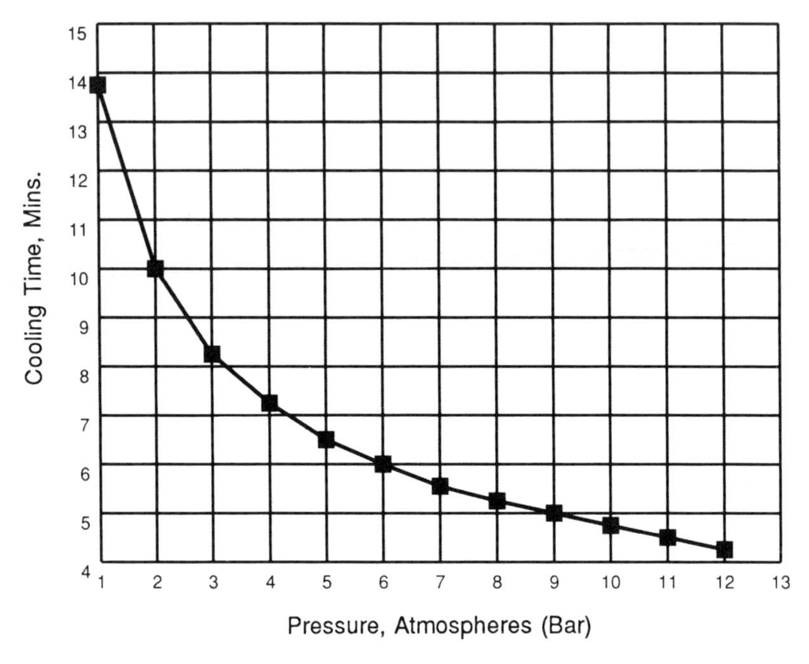

Fig 6.6 Cooling time vs gas pressure for vacuum heat-treating furnace

large workloads of high-speed M2 tool steel gear cutter hobs. Faster cooling can reduce the overall process cycle time sufficiently to justify this added cost.

Slow-Cooling Processes

Not all workloads require quick cooling or quenching to ambient temperature. Often brazing work is allowed to cool slowly in vacuum to 55 or 110 °C (100 or 200 °F) below brazing temperature, to allow the braze to solidify, and is then backfilled and gas-quenched. Annealing of tool steels, complex weldments, and magnetic metals, and processing of certain powdered metals, all require a slow cooling rate, on the order of 14 °C (25 °F) per hour, after thermal soaking. In this type of program cycle, the furnace hot zone and work remain under vacuum, and the power supply stays on during a controlled power-down regulated by the furnace temperature programmer/controller. When the work reaches a suitably low temperature, the furnace may be backfilled to atmospheric pressure. These cycles can require many hours of production time.

Where programmed power-on cooling rates are not fast enough but gas-cooling rates are too fast, a combination of power-on cooling and gas cooling can be used with a two-channel controller/programmer to achieve the requisite cooling rate. One channel controls the furnace power supply, while the second channel controls the throttle position of the gas control valve. Programming the two-channel controller to meet the cooling rate requirements can be challenging (Ref 7).

References

1) W.R. Jones, Pumping and the Vacuum Furnace, *Heat Treating*, July 1986
2) W.R. Jones, Vacuum—Another Atmosphere?, *Heat Treating*, October 1986
3) W.R. Jones, Hot Zone Contamination, *Heat Treating*, January 1989
4) "Televac Vacuum Gauge, Model MM 200," The Fredericks Co., Huntingdon Valley, PA
5) W.R. Jones, High-Velocity Gas Flow Seen As Key to Rapid Quench, *Heat Treating*, September 1985
6) W.W. Hoke, II, High-Pressure Cooling Performances in Vacuum Heat Treating Furnaces as Analyzed by New Method, *Industrial Heating*, March 1991
7) W.W. Hoke, II, Controlled Cooling Rates for Steel Plates Using Work Thermocouples in a Vacuum Furnace, *Industrial Heating*, September 1990

• 7 •

Vacuum Instrumentation and Control

David Holkeboer, Leybold Inficon, Inc.

Introduction

The purpose of vacuum measurement instrumentation is to measure the amount of gas atmosphere remaining in a vacuum vessel. For some vacuum furnace processes it is sufficient to determine that the density of the atmosphere is sufficiently low. In other cases, for example rapid cooling, a suitable level of gas pressure is required. In still other cases the nature of the atmosphere is also important. Finally, when the required vacuum conditions cannot be readily achieved, vacuum instruments are helpful in locating the source of trouble.

Vacuum levels are traditionally measured in units of pressure, although the forces resulting from the pressure in most vacuum systems are negligible and often not directly measurable. In this discussion, pressure will mean absolute pressure, where zero is the pressure in a wholly empty vessel. "Absolute pressure" is distinguished from "gage pressure," which is the difference between atmospheric pressure and the absolute pressure in a vessel. Absolute pressure is a measure of the amount of the gas atmosphere in a vacuum vessel and is a better indicator of the effect of the atmosphere on the process than gage pressure.

The units for pressure measurement in the United States are the torr (= 1 mm of mercury) and the millitorr (= 1 micron of mercury); 760 torr = 1 standard U.S. atm. Other common units are the millibar (.001 atm) and the Pascal (.01 millibars). One millibar is 0.76 torr, and for many purposes can be taken as approximately 1 torr.

This chapter will consider vacuum instruments as follows:

- Mechanical and electrical vacuum gages, for measurement of pressure
- Leak detectors, for vacuum system maintenance
- Residual gas analyzers, for gas component measurement

Mechanical Vacuum Gages

Mechanical vacuum gages measure the actual force exerted by the gas in the vacuum vessel. Since this force becomes very small at low absolute pressures, the mechanical gages cannot measure very high vacuum levels.

Mechanical Bourdon-Tube and Diaphragm Gages

These are the simplest vacuum gages. Many of them register their readings on a mechanical dial, but some have electronic readouts. Often these gages are an exception to the principle that absolute pressure is the quantity to be measured; they frequently indicate gage pressure, often in inches of mercury.

Operating Principle. Bourdon gages contain a thin-walled metal tube flattened and bent into a circular arc. Increasing gas pressure in the tube straightens the arc out, and when pressure decreases the tube returns to its bent shape. The end of the tube is linked to a lever and gear mechanism which moves a pointer on the dial face. Diaphragm gages have a flat metal diaphragm exposed on one side to ambient pressure and on the other to the vacuum space. The movement of the diaphragm in response to vacuum level changes is linked to a lever and gear mechanism which moves a pointer on the dial face. To convert either type to measure absolute pressure, the case containing the Bourdon tube or diaphragm assembly must be evacuated and sealed.

Operating Characteristics. Pressure measurements by both of these types of mechanical gages are independent of the kind of gas being measured. If the gage does not measure absolute pressure, all measurements are relative to the ambient atmospheric pressure, which is affected by altitude and weather conditions. The pressure levels indicated by the gage are thus subject to fluctuations in the ambient barometric pressure.

Capacitance Manometer Vacuum Gages

Capacitance manometers are the most sensitive of the mechanical diaphragm vacuum gages used for vacuum measurement. They are made in

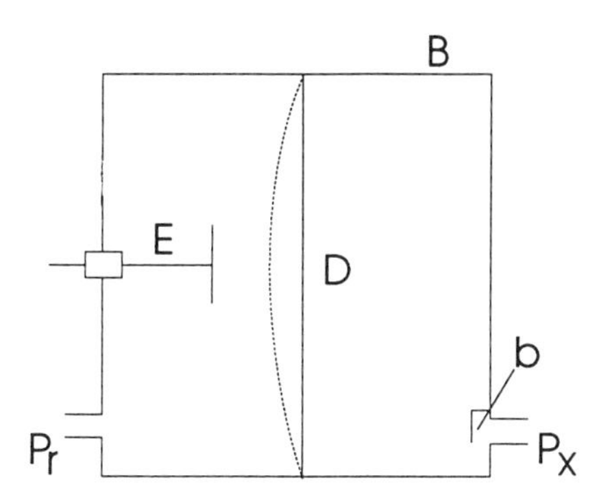

Fig 7.1 Capacitance manometer. **E**, electrode; **D**, elastic diaphragm; **B**, leak-tight box; $\mathbf{P_x}$, pressure to be measured; $\mathbf{P_r}$, reference pressure; **b**, baffle.

several models, ranging from 1000 torr to 1 torr full-scale reading. To measure lower pressures it is common to amplify the gage output by a factor up to 10,000 or 100,000. For example, when using a 10 torr sensor to measure pressures in the range from 1×10^{-4} torr to a 1×10^{-3} torr, the readings are being multiplied (electronically) by 10,000. Capacitance manometers measure pressure independent of gas type and expose only stainless steel and Inconel to the gas being measured.

Operating Principle. Capacitance manometers measure pressure by detecting the deflection of a thin metal diaphragm exposed on one side to the pressure being measured and on the other to a reference pressure (Fig 7.1). Motion of the diaphragm is detected by measuring the change in capacitance between the diaphragm and a stationary electrode. The measuring electronics are normally integrated into the sensing unit. These devices are sensitive enough to detect diaphragm displacements of less than the diameter of one molecule. Sensing units are typically individually linearized and calibrated in the manufacturing process. Capacitance manometers are available as either absolute or differential gages. Absolute pressure gages, which have one side of the diaphragm evacuated and sealed, are more convenient for most purposes. Differential gages have a connection port on each side of the diaphragm and are sometimes preferred for metrology.

Operating Characteristics. Capacitance manometers are capable of good accuracy, better than 1% of reading, when used near their full-scale values. However, when reading a small fraction of full-scale, several significant sources of error may reduce the accuracy greatly:

1) They require zero adjustments, accomplished by pumping them down to a negligible pressure compared to the measurement range and resetting the indicator to zero. Manufacturers' accuracy specifications all assume this adjustment has been made. Units with 1 or 10 torr full-scale ranges must be re-zeroed any time they are exposed to atmospheric pressure. For such a unit the recommended solution is to install a valve between it and the vacuum vessel so that it is not subjected to pressures greater than its full-scale rating.

2) They are sensitive to temperature changes. For example, with a temperature coefficient of 50 ppm/°C, a 1-torr full scale sensor used to measure 1×10^{-3} torr develops a 5% error when the temperature changes 1 °C (1.8 °F). At 1×10^{-4} torr the error is 50%. To minimize this effect, some units are fitted with a heater and thermostat to maintain them at a constant temperature, typically 45 °C (113 °F).
3) They are moderately sensitive to vibration. Heavy vibration or shock can shift their zero point. More moderate vibration produces an erratic, noisy output.
4) They are subject to thermal transpiration pressure errors. At pressures less than 1 torr, the pressures in two connected vacuum vessels at different temperatures are unequal due to an effect known as thermal transpiration. In fully developed thermal transpiration, which occurs at pressures less than 1×10^{-3} torr, the ratio of the pressures is equal to the square root of the ratio of the absolute temperatures of the two vessels. For gages that are heated for stabilization to 45 °C (113 °F) this error amounts to about 2.5% of the reading, which is probably not significant in most cases.

Electronic Vacuum Gages

There are three basic types of electronic vacuum gages: (1) thermal, (2) cold-cathode, and (3) hot-cathode.

Thermal Vacuum Gages

Thermal vacuum gages measure pressure by detecting how well the gas transfers heat. Their principle operating range is 1 torr to 1×10^{-3} torr, but some can read pressures up to 1 atm. The common types of thermal gages are the Pirani gage and the thermocouple gage.

Operating Principle. The sensing element in a thermal vacuum gage is a fine metal wire which is heated to approximately 100 °C (212 °F) by passing an electric current through it. In thermocouple gages, this temperature is measured with a thermocouple and the reading is correlated with gas pressure. In Pirani gages, the wire temperature is measured by reading its electrical resistance. In some Pirani gages, resistance of the wire is correlated with the pressure. More commonly, these gages use an electronic feedback circuit to maintain a constant resistance, signifying constant temperature, and correlate heating power with gas pressure.

In principle, the thermal conductivity of a gas is independent of the pressure, and the heat loss from a heated wire in still air is constant.

However, at very low pressures, where the mean free path of gas molecules is similar to or larger than the diameter of the wire, the rate of heat loss is proportional to the rate of arrival of molecules at the surface of the wire. Thus, as the vacuum improves, the wire temperature rises, or the power needed for constant temperature decreases. The low pressure limit for this kind of gage is reached when radiation heat loss from the wire and conduction to its supports become the dominant heat-loss mechanisms. In some instruments of this type, the high-pressure range is extended up to 1 atm by the detection of convection current heat loss created in the gage by the heated wire. Because the convection effect is only weakly dependent on pressure, the high-pressure readings have correspondingly limited accuracy.

Operating Characteristics. Thermal conductivity gages are sensitive to ambient temperature changes and gas type. Those using the convection principle are also sensitive to sensor orientation. Many designs include temperature compensation in the gage. The effect of changing gas types may be large. Using this kind of gage to control vacuum chamber backfill to 1 atm is very dangerous if the type of gas is not correct, because significant overpressures can occur. A better method is to use a mechanical pressure switch. Thermal conductivity gages are commonly used to indicate the pressure on roughing and backing pump systems and as interlock sensors to control the operation of hot-cathode gages and other instruments.

Cold-Cathode Gages

Cold-cathode ionization gages are a type of high-vacuum gage used to measure pressures of approximately 1×10^{-3} torr or less. They do this by ionizing the gas molecules and measuring the resultant current. They are also called Penning, Phillips, or magnetron gages.

Operating Principle. Operation of a cold-cathode gage is shown in Fig 7.2. This gage contains two electrodes, a cathode or electron source, and an anode or electron collector. The anode is cylindrical or ring-shaped and it operates at a potential of 1000 to 3000 v positive with respect to the cathode. The gage operates in a magnetic field of approximately 1000 gauss (0.1 tesla), which serves to trap electrons which cannot cross the field to travel from the cathode to the anode. Thus the gage contains a cloud of trapped electrons. Ionization occurs when a trapped electron hits a gas molecule hard enough to make it lose one or more electrons or to make a complex molecule break into several fragments, yielding one or more positive ions. These ions are attracted to the cathode, and being relatively massive, are not trapped by the magnetic field. Measurement of the current of ions reaching the cathode indicate the pressure. The electron cloud is replenished by secondary elec-

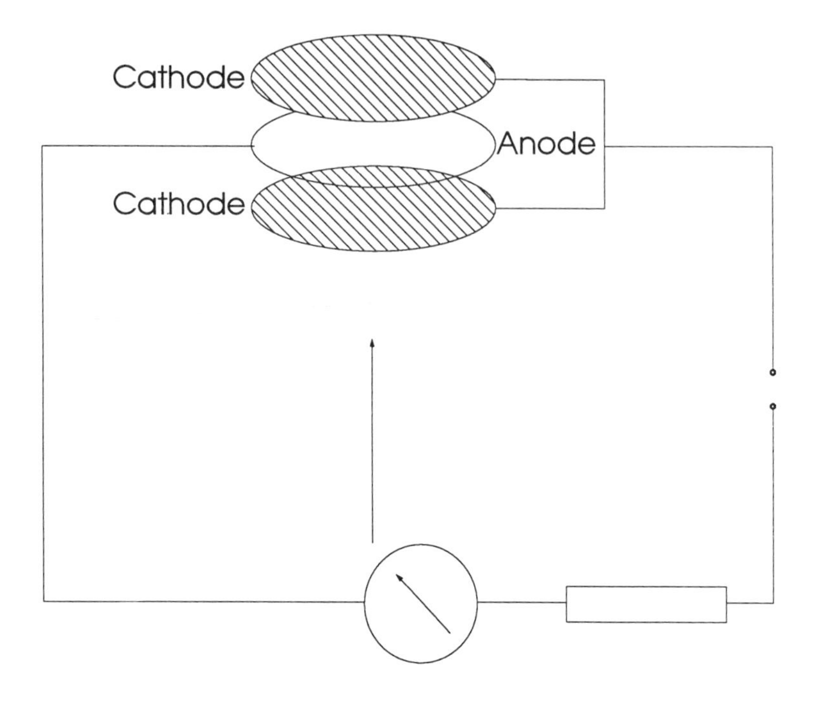

Fig 7.2 Cold-cathode ionization gage

trons emitted from the cathode when gas ions strike it, as well as by electrons from the ionization process itself. Thus, the electron cloud remains in a stable, saturated condition. In fact, a current of electrons escaping to the anode equals the ion current reaching the cathode, and either can be used to indicate the pressure.

Cold-cathode gages are made in several configurations, called Penning or Phillips gages, magnetron gages, and inverted magnetron gages. They differ in their ability to hold a stable cloud of trapped electrons and thus in their performance. Penning or Phillips gages typically have a ring-shaped anode and two flat-plate cathodes. They are the least costly and often the least stable. Magnetron gages have concentric cylindrical anodes and cathodes, with cathode end plates to prevent the escape of electrons in the axial direction. A magnetron gage has its cathode inside the anode, while an inverted magnetron gage has it outside (Fig 7.3).

Operating Characteristics. Compared with hot-cathode ionization gages, cold-cathode gages have simpler, more rugged sensors and less costly

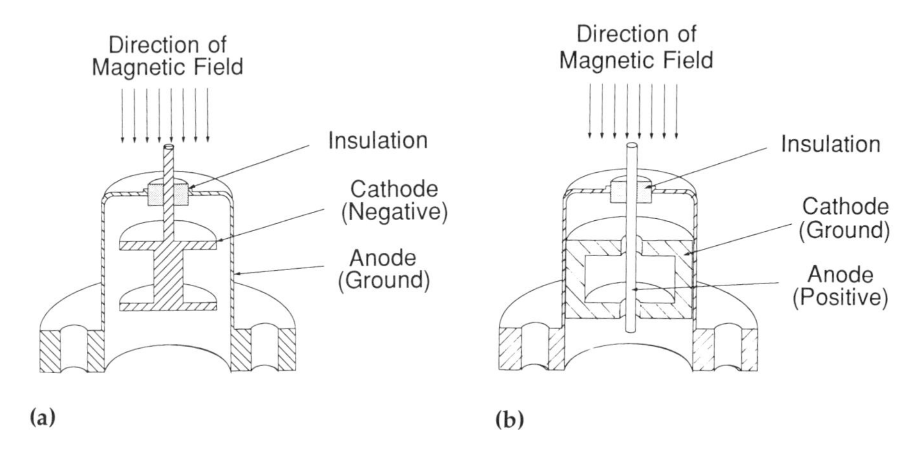

Fig 7.3 (a) Magnetron and **(b)** inverted magnetron gages

electronic control systems. Because they can be dismantled and cleaned and have no filament to burn out, they are often used in applications with high contamination levels. Starting a cold-cathode gage requires an initial ionization from a stray electron, and cold-cathode gages are often hard to start in high vacuums, where the concentration of gas molecules is low. Gages that trap electrons poorly may allow stray electrons to escape before an ion is formed. A variety of starting aids, such as sharply pointed electrodes and dual-ionizing volumes have been used by various manufacturers. Sharp points, which yield electrons due to field emission work well when new, but the points tend to become eroded during use and effectiveness decreases. Deposits built up in the gage often affect starting, but not always adversely. Thus, the same model gage may start reliably in one application and not in another.

It is common for cold-cathode gages to exhibit instability at some pressures, accompanied by sudden changes in output that do not correspond to real pressure changes. This seems to be more common in the simple Penning gages than in the magnetrons, and such effects are thought to be caused by changes in or oscillations of the electron cloud inside the gage. Cold-cathode gages are similar in principle to the magnetron oscillators used to generate microwaves; the gages are not intended to oscillate, but they sometimes do, with the result that their measured output current changes significantly.

The relationship between pressure and gage output is dependent on the type of gas being detected, in a manner similar to that of hot-cathode gages. The following section contains a table of typical gage sensitivities.

Hot-Cathode Gages

Hot-cathode ionization gages measure pressure by ionizing molecules of the gas and measuring the resulting current. The common type of hot-cathode gage is also known as the Bayard-Alpert gage.

Operating Principle. Figure 7.4 diagrams a hot-cathode ionization gage. It contains three elements: a filament, a cylindrical screen anode (usually called the grid), and an ion collector. The filament is typically about 30 v above ground, the grid at 180 v, and the collector at ground potential. Electrons to ionize the gas molecules are supplied by heating the filament to a suitable temperature to achieve the desired electron emission current. The electrons are accelerated toward the grid by its positive potential, but the grid is quite transparent, so that many of them pass through it, some several times, before they are eventually collected. The electrons strike gas molecules with sufficient energy to knock electrons out of them, creating positive ions. Complex molecules often fragment in this process and can yield more than one ion. The positive ions are repelled by the positive grid and are attracted to the ion collector, where the resulting current is measured. This current is correlated with the pressure.

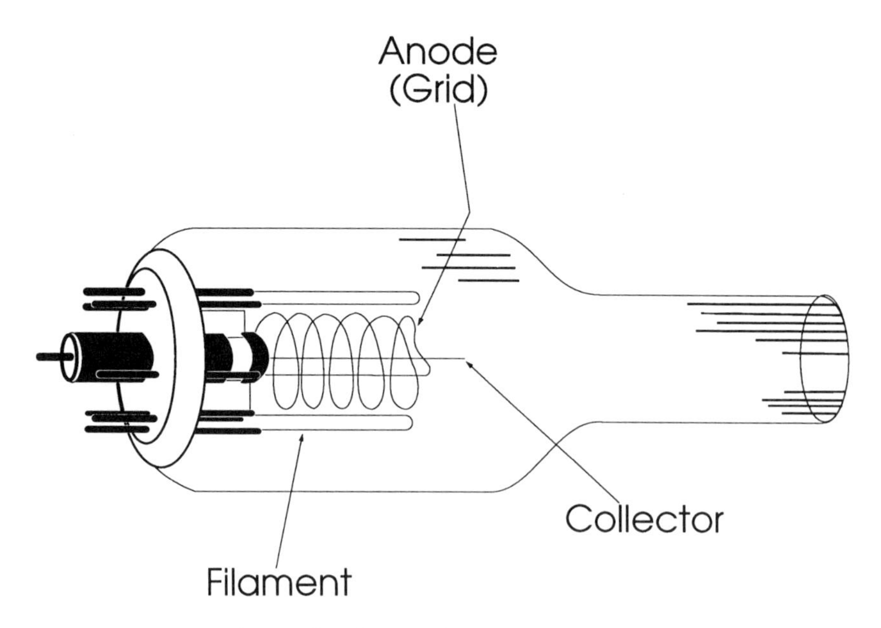

Fig 7.4 Tubulated (glass enclosed) ionization gage

Hot-cathode ionization gages can be used only in high vacuums, with pressures less than approximately 1×10^{-3} torr, where electrons and ions can move through the gas atmosphere independently. At these pressures, gage output is determined by the magnitude and distribution of the electron emission current, the concentration of gas molecules, and the ionizing efficiency for gas molecules. Hot-cathode gages are normally operated by a dedicated electronic unit which supplies regulated voltages to the electrodes, controls the electron emission current, and amplifies the gage output for display in pressure units.

Operating Characteristics. To prolong their service life, the filaments of hot-cathode gages should be kept turned off when the vacuum level is not suitable for gage operation. Two types of filaments are available: Tungsten filaments are inexpensive and they resist contamination, but they are subject to burn-out if turned on when there is insufficient vacuum. A secondary sensor, such as a Pirani or thermocouple vacuum gage is often employed to prevent operation of the filament at unsuitable vacuum levels. "Non-burn-out" filaments are made of iridium with a coating of thorium oxide or yttrium oxide. Although their coatings are attacked by chlorides and fluorides, they are otherwise quite durable.

The relationship of gage output to absolute pressure is called sensitivity and depends on the type of gas being measured. Manufacturers normally design their gages for nitrogen gas. Table 7.1 shows typical pressure correction factors for pure gases relative to nitrogen. Note that a wide variation in sensitivities is reported in the literature, so that the data in the table should not be regarded as precise.

Hot-cathode gages are not normally calibrated by their manufacturers. Manufacturing tolerances may cause new gages to be in error by ±30%, or

Table 7.1 Typical ionization gage sensitivity relative to nitrogen standard

Gas	Sensitivity
Air	1.00
Argon	1.20
Carbon Dioxide	1.40
Carbon Monoxide	1.05
Helium	0.14
Hydrocarbons	6.0-7.0
Hydrogen	0.45
Methane	1.60
Nitrogen	1.00
Oxygen	1.00
Water	1.00

more. Gage sensitivity changes during use, particularly if the sensor becomes contaminated. For reliable measurement therefore, hot-cathode gages must be recalibrated at suitable intervals. Calibration for process certification is carried out by comparison with a recognized measurement transfer standard, which must be traceable to the National Institute for Standards and Technology (NIST), formerly the National Bureau of Standards, in Gaithersburg, Maryland. This calibration service, which requires special equipment, is available from several commercial sources, including most gage manufacturers.

At pressures greater than about 1×10^{-4} torr, hot-cathode gages tend to become non-linear. Some gages can provide an extended linear range by reducing the electron emission current and adjusting the output display accordingly. On first entering the non-linear range, the gage output does not rise as rapidly as the pressure. If the pressure increases farther, the output levels off at a maximum value, and eventually drops. In this saturation region, hot-cathode gage readings are grossly inaccurate, although they may appear plausible.

Hot-cathode ionization gages are built in two configurations: glass-enclosed (tubulated) gages, and open (nude) gages (see Fig 7.4 and 7.5).

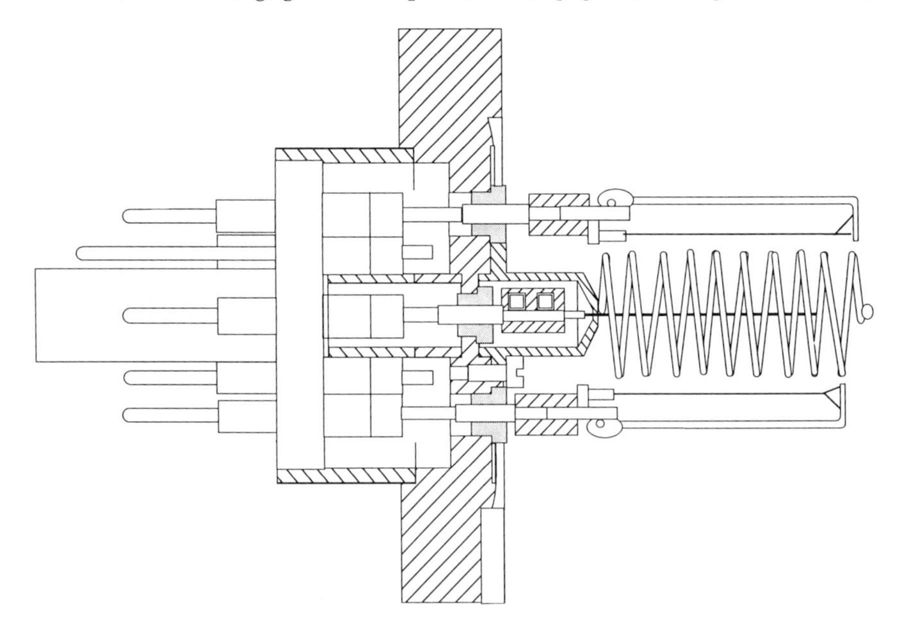

Fig 7.5 Open (nude) gage

Tubulated gages cost less than nude gages, but the filaments of nude gages are replaceable. At high pressures, tubulated gages are subject to a change in wall potential which can change the gage reading by as much as 50%. Some manufacturers offer tubulated gages with metallized walls, which effectively prevent this behavior. For ultra-high-vacuum situations, tubulation of the gage can cause pressure differences between the gage and the vacuum work space. Except for these two considerations, there is little difference in performance between the two types. Tubulated gages can present an implosion hazard if accidentally broken while under vacuum, and they should always be enclosed by a protective guard.

Helium Leak Detector

Anyone operating a vacuum facility must deal with leaks which allow air, or sometimes cooling water, to escape into the vacuum zone. Leaks can make it impossible to attain the required vacuum, or even if the vacuum level is reached, can spoil the product, by allowing oxidation of the surface, for example. Leak detectors are useful for finding leaks so that they can be repaired, and for proving that leaks are not present before starting a critical process. The latter is relatively the more difficult task, because the leak detector can detect only those leaks that occur in areas where the user suspects they exist. And the user must be sure that the leak detector has been correctly used to test every possible leak site.

Operating Principle

The helium leak detector consists of a special mass spectrometer coupled to a small vacuum pumping system. The mass spectrometer is designed for the detection of helium only and is optimized for very high sensitivity. The pumping system provides the necessary vacuum for operation of the mass spectrometer and supports the gas flow necessary to get helium into and out of the spectrometer. The limit of detection for commercial helium leak detectors varies from about 1×10^{-9} to 1×10^{-12} torr-liters/s. However, when a large amount of other gases, such as air or water vapor, is present, its sensitivity is limited to about 1 to 10 ppm.

For detecting leaks in a vacuum vessel or system, the helium leak detector is connected to the vacuum space. Helium gas is applied to the atmospheric side of any areas suspected of having leaks. The leak detector signals leakage of helium into the vacuum space, permitting the location and size of the leak to be determined.

Unlike the RGA mass spectrometers described later in this chapter, the helium leak detector mass spectrometer is a typically a magnetic sector instrument. It has a hot-cathode ion source similar to a hot cathode ionization gage. Ions formed in it are accelerated and focussed into a beam that is projected through a slit into a magnetic field. In this field, individual ions follow curved paths, the curvature of which is dependent on the ion mass. The mass spectrometer is adjusted so that only helium ions, with a mass of 4 atomic mass units (amu), have the proper trajectory to reach the entrance slit of the detector, where the ion current is measured.

A leak detector pumping system consists of a high vacuum pump and a backing pump. The high-vacuum pump for older helium leak detectors was an oil diffusion pump, and these systems often required liquid-nitrogen-cooled vapor traps to prevent spectrometer contamination and improve pumping performance. Most newer units have turbomolecular pumps, which are more convenient to use. In these systems the pumps, sample

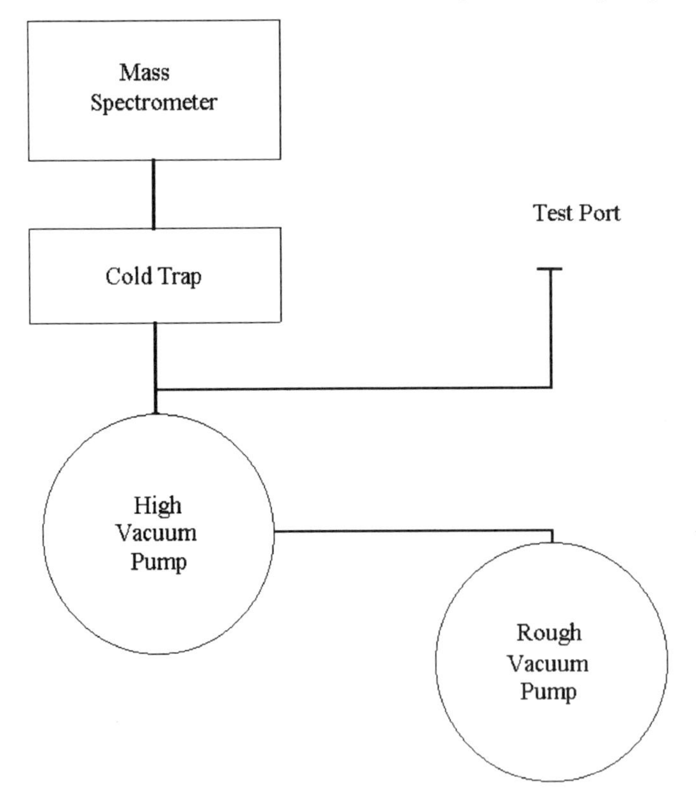

Fig 7.6 Direct sample leak detector system

valves, and mass spectrometer are operated by an interlocked control system for convenience and safety.

The pumping system may provide for direct sample input or counterflow sample input. In a direct sample system, as shown in Fig 7.6, gas flow from the device under test goes directly into the mass spectrometer region and is subsequently pumped away by the high vacuum pump and backing pump. The pumping speed is reduced by an orifice to about 1 liter/s, so that the concentration of helium will be high enough for detection even if the leak is very small. In a counterflow system, as shown in Fig 7.7, the gas sample enters the backing system at the discharge side of the high-vacuum pump. Helium gas, consisting of relatively light molecules, diffuses against the counterflow pumping action of the high vacuum pump and reaches the mass spectrometer. At the same time heavier gases, such as water vapor and

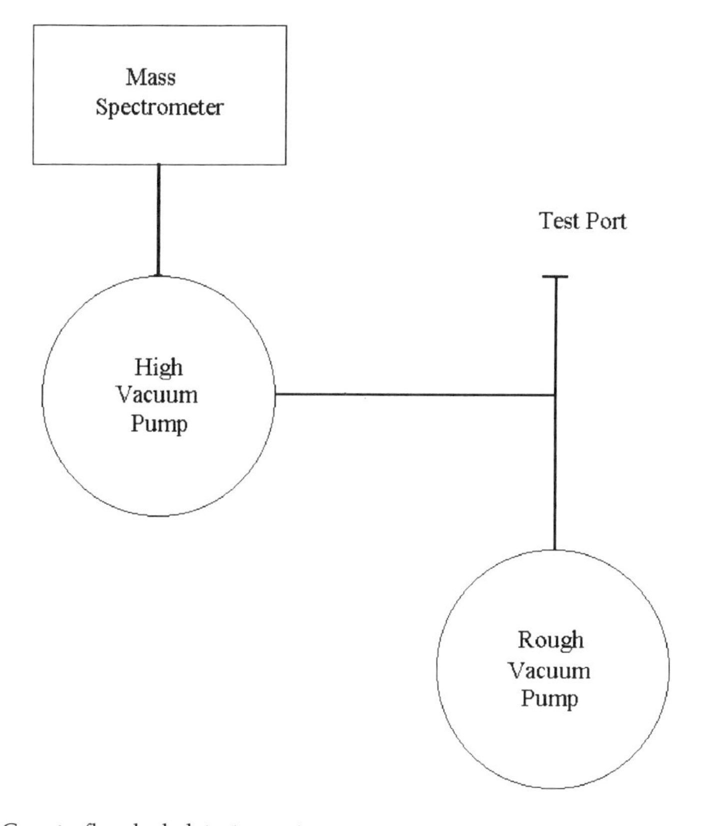

Fig 7.7 Counterflow leak detector system

air, are prevented from entering the mass spectrometer in excessive quantities. Direct-input leak detectors have the advantage of being more sensitive by a factor of 100 or so. However, the mass spectrometer is also exposed to the full amount of water vapor, air, and whatever contaminants may be in the gas being sampled.

Operating Characteristics

Because the pumping speed of a helium leak detector is relatively low (about 1 liter/s), by itself it cannot evacuate a significant volume such as the chamber of a vacuum furnace, and the pumping system of the furnace must be operated to produce a suitable vacuum for the leak detector to sample.

One possible configuration is to connect the leak detector to a suitable port on the vessel, as shown in Fig 7.8. In making this connection, the location of the port is not particularly important, but the connection should be kept short, large in diameter, and metallic. For example, a 19 mm (3/4 in.)

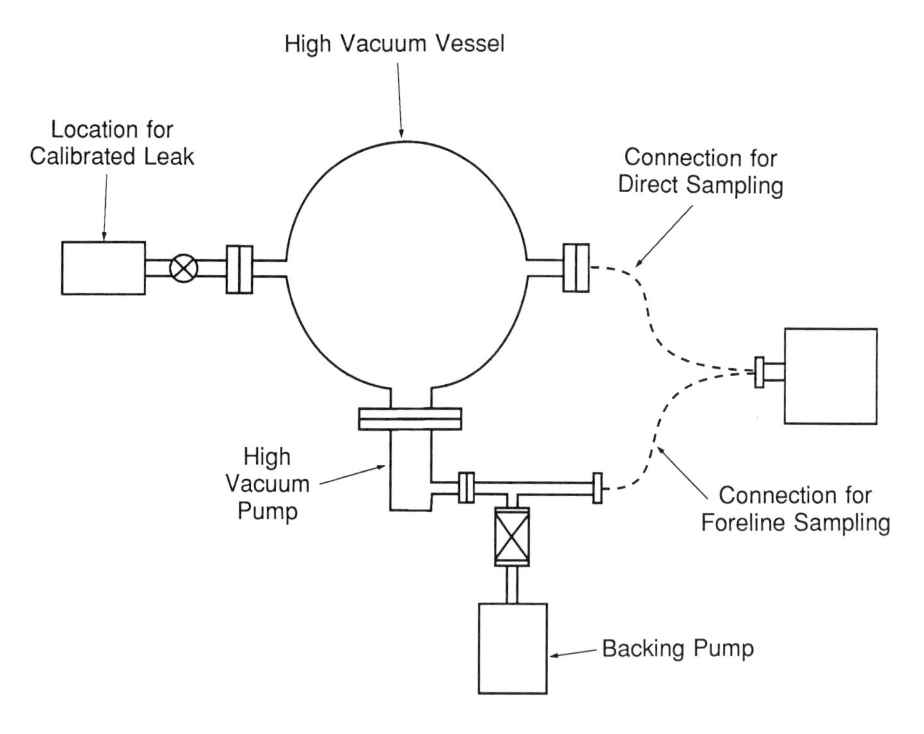

Fig 7.8 High vacuum pump and backing pump combination system

ID hose 1000 mm (40 in.) long will reduce leak detector sensitivity by 50%. If the leak detector is connected to a port on the vacuum vessel, the gas flow will be divided between the leak detector and the main pumping system in proportion to the pumping capacities of each. For example, if the main pumping system has a speed of 1000 liters/s and the leak detector has a speed of 1 liter/s, sensitivity of the leak detector is reduced by a factor of 1001.

There is also a time lag, determined by the volume of the vacuum vessel and the total pumping speed. The time constant of a vacuum system is determined by dividing the volume by the pumping speed. For example, a 250-liter vessel with a 1000 liter/s pumping system has a time constant of 0.25 s. The same vessel with a 25 liter/s pumping system has a time constant of 10 s. A change in the level of a non-condensible gas such as helium, is 90% complete in 2.3 time constants, and 99% complete in 4.6 time constants. Thus, with the larger pumping system in this example, the full value of a leak will be observable within less than 1 s after helium is applied to it. With the smaller pump, it will take 23 s to see 90% of the full leak signal. In the latter case, it is important to apply helium for a sufficiently long time period to each suspected leak area to be sure of detecting any leak that may exist there.

A second method used for systems where the vacuum vessel is pumped by a high-vacuum pump and backing pump combination, is shown in Fig 7.8. In this case, the leak detector is connected into the foreline of the backing pump. The advantage of this arrangement over the preceding one is that its division of gas flow offers higher sensitivity, while it retains the fast response of a high-speed pumping system. These performance improvements result because the high-vacuum pump serves to collect the gases, including helium, from the large vacuum vessel and deliver them in compressed form to the leak detector and backing pump. Since the backing pump has a much lower pumping speed than the high-vacuum pump, it does not take as much of the gas away from the leak detector, which makes the sensitivity loss from the divided gas flow much less. The disadvantage of this arrangement is that a leak detector at this location is often more subject to contamination by pump oil.

To ensure reliable operation, leak detectors must be calibrated immediately before use. For this purpose, calibrated helium leaks are available in a variety of sizes. Some leak detectors are furnished with built-in calibrated leaks which may be operated automatically. While these are useful for instrument maintenance and tuneup, they tell little about the effective sensitivity when connected to a vacuum system. A much more reliable way of ensuring that a vacuum system is free of leaks is to install a calibrated helium leak that can be turned on and off at will, on the vacuum vessel, as shown in Fig 7.8. This will permit verification of both the magnitude and time delay of the leak detector, to demonstrate that it will detect any real leaks that might be present.

Quadrupole Residual Gas Analyzer

An RGA (Residual Gas Analyzer) is a small mass spectrometer designed for direct insertion in a high vacuum vessel for the purpose of indicating the constituents in the residual gas remaining in the vessel after (or during) evacuation. The term "quadrupole" refers to the four metal rods that constitute its ion beam filter.

Operating Principle

The RGA identifies gases by displaying the characteristic peak or family of peaks that results from ionizing these gases. For example, helium yields one peak at 4 amu; nitrogen yields two peaks, the larger one at 28 amu and the smaller at 14 amu; while water shows 3 peaks at 16, 17, and 18 amu (see Fig 7.9). This identification can be much more useful than a simple pressure measurement for vacuum system trouble shooting and process monitoring. Unfortunately, identification is not always unequivocal; carbon monoxide has a 28 amu major peak, and carbon dioxide also has a large 28 amu peak.

In addition to identifying the gas constituents in this manner, in a non-quantitative way the amplitude of the peaks indicates the amount of the gas constituent present. Some instruments report RGA output readings in pressure units (torr or mbars) but unless the gases are known and the proper calibration factors entered, these readings can be significantly in error.

An RGA sensing device comprises an ionizer, an ion selector or filter, and an ion detector, per Fig 7.10. To detect the gas molecules, they are electrically charged (ionized) in the ionizer, using electrons emitted from a hot filament. The ionizer operates in similar manner to a hot-cathode ionization gage, except that instead of being collected immediately the ions are accelerated and focussed into the ion selector. The ion selector commonly used today is the quadrupole mass filter, although some instruments use a magnet as a mass filter. The ion selector separates the ions in the mixed beam generated by the ionizer according to ion mass. Separation is accomplished by removing all those ions which do not have the mass to which the selector is set.

The quadrupole mass filter consists of four parallel metal rods which form an ion beam passage from the ionizer to the detector, as shown in Fig 7.9. The rods are connected in pairs to a control circuit which provides both radio frequency (rf) and direct current (dc) voltages. Positive ions in the

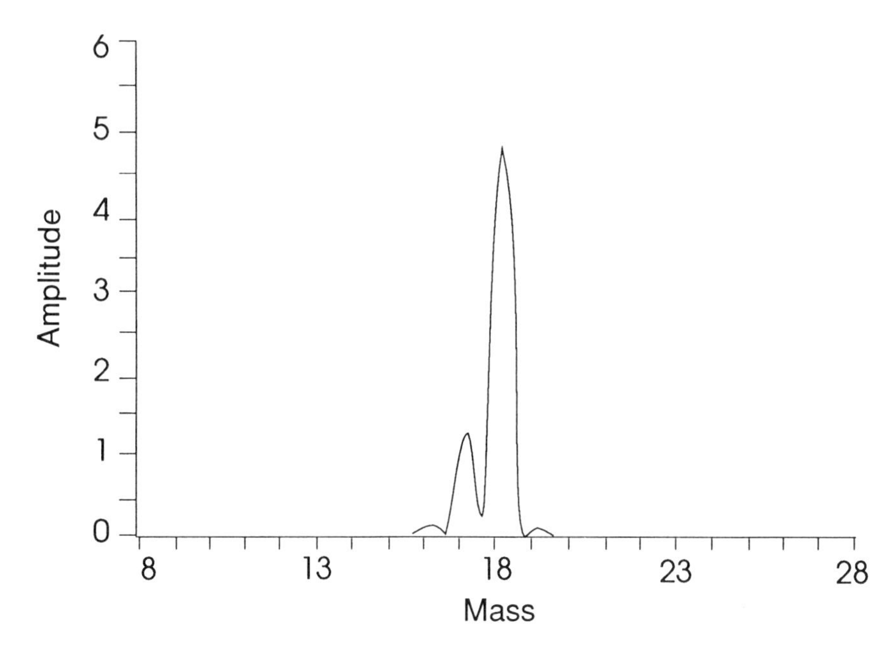

Fig 7.9 Analog/water spectra display

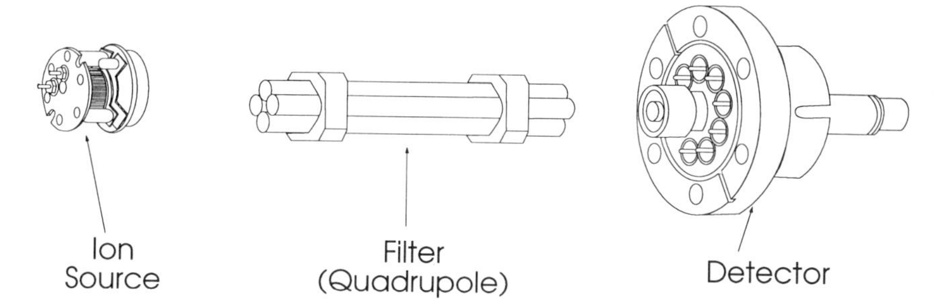

Fig 7.10 Quadrupole RGA mass filter

beam are attracted to the negative poles of the filter, and they can thus be pulled out of the beam and prevented from being detected at the collector.

The function of the rf component of the electrical field is to eject low-mass ions and the function of the dc field is to eject high-mass ions. By varying the rf and dc field intensity, it is possible to eject all ions heavier and lighter than one particular mass, so that the current of ions of one given mass can be measured. In effect, the filter is tuned to that mass.

Those ions not removed by the filter reach the detector, where the resulting current is collected and measured. The detector may be a simple Faraday cup collector or an electron multiplier. An electron multiplier acts as a low-noise, high-gain amplifier for very small ion current signals.

If the rf and dc fields are swept from low to high voltage, and the intensity of the ion current is measured during the sweep, a mass spectrum can be recorded. As each mass in the range is brought into tune, the current rises, reaches a maximum, and then falls. This is called a peak; peaks have a height (expressed in amperes) determined by the number of ions being detected, and a width determined by the resolving power of the filter. Identification of the peaks and measurement of their height is basic to the use of the RGA.

It is not necessary to sweep the entire mass range to get useful data. An RGA can sweep a selected part of the mass range, monitor a single peak or several selected peaks, and provide control outputs based on peak heights.

Operating Characteristics

Quadrupole sensors are relatively sensitive to the characteristics of the electronics unit driving them, because the mass filter behavior depends strongly on the accuracy of the rf and dc drive voltages relative to the dimensions of the quadrupole rod assembly. The electronics unit must be adjusted or calibrated to match the sensor. The coupling between the rf and dc generator and the quadrupole is part of the tuning circuit, and it must not be changed after the electronic unit has been calibrated.

Vacuum conditions for quadrupole RGA operation are similar to but somewhat more restrictive than those for hot-cathode ionization gages. The maximum useful operating pressure is typically less than 1×10^{-4} torr. RGAs which have electron multiplier detectors are limited to even lower maximum pressures, usually less than 1×10^{-5} torr, to avoid damage to the multiplier. Contaminated vacuum systems may cause serious performance degradation if the quadrupole filter rods become coated by contaminating materials. The most common contaminants are greases and oils. Contaminated sensors often show distorted peak shapes. Another characteristic is peak amplitudes which drift with time when a single peak is observed. Contaminated sensors usually have to be dismantled and cleaned.

Quadrupole RGA instruments are made with various mass ranges; the lowest mass is normally 1 or 2 amu and the highest mass may be 100, 200, 300, or 400 amu. For vacuum furnace applications, mass ranges greater than 100 amu will seldom be needed. Within its mass range, the RGA will measure and display the amplitudes of ion current peaks for each mass of ion produced by the ionizer. The range of currents that can be detected is quite large, often 6 or 7 decades. Most instruments are designed so that their scanning speed is automatically adjusted to match the amplification used, so that the output does not become excessively noisy. When the ion currents are large, a 100 amu spectrum can be scanned in less than 1 s; at maximum amplification, the same scan can require several minutes.

Identification of the gas species represented by the array of ion peaks in the mass spectrum is normally left to the user. Fortunately, most of the common gases that are important to the vacuum furnace operator are readily distinguishable by observing one or more prominent peaks. These, with their amu numbers, include hydrogen, 2; helium, 4; water vapor, 18; oxygen, 32; argon, 40; and carbon dioxide, 44. Nitrogen and carbon monoxide both have a mass of 28 amu and are individually distinguishable only by their fragment ions. On the other hand, oils and greases tend to produce large numbers of similar peaks in the spectrum, which are readily observable but not easy to assign to a source. Typical peaks from many hydrocarbons, which are fragments of larger molecules, are found at amu numbers 27, 29, 39, 41, 43, 55, 57, 67, 69, 71.

In addition to identifying them, it is often desirable to know the quantities of the different gas components in a vacuum vessel or furnace. Unfortunately, the partial pressures of individual gases cannot simply be read from the spectrum, like a vacuum gage.

Limitations

The principal limitations of the quadrupole RGA are:

1) Ionization of gas molecules is its basis for detection, as in the ionization gage. Gas species differ in the total number of ions produced and division of this ion yield among the various masses.
2) The transmission of its ion filter is a strong function of ion mass. For masses greater than 20 amu or so, the transmission is approximately inversely proportional to mass. At very low masses such as 1, 2, and 4 amu, the transmission also tends to decrease, usually in an unpredictable way. This means that the partial pressure of species of high molecular weight tends to be strongly under-represented in its output spectrum.

3) Its sensitivity is the ratio of the peak height (in amps) to the pressure of a standard gas, usually nitrogen. Manufacturers usually quote a minimum standard for new production units, but actual sensitivity may be significantly higher. For electron multiplier sensors, sensitivity varies widely (by a factor of 10,000 or more) with changes of multiplier voltage. Sensitivity is also affected by rf and dc drive calibration or other changes in drive performance.
4) Quadrupoles become non-linear as the upper pressure limit is approached. Some designs are much more linear than others. Non-linearity may be more severe with heavy gases than with light. Sensor condition and operating parameters can also effect linearity. Concentration measurement should be confined to the linear range.

Some instruments provide for the entry of sensitivity and gas type constants so that true partial pressures will be displayed. In other instruments a simple scaling factor is applied to output currents and the results are labelled in pressure units. This latter scheme can result in large errors and should not be relied upon.

Applications

The principal uses of the quadrupole RGA in vacuum furnace practice are leak detection and partial-pressure measurements.

Leak Detection. Leak detection is a common and valuable RGA application. The RGA has several advantages over the helium leak detector in finding leaks in a high-vacuum system. First, the RGA spectrum can distinguish leakage from other vacuum system problems, by detecting significant amounts of leaking matter, such as air, water, argon, etc. Second, the RGA can use many gases other than helium as leak location tracers. The disadvantages of using the RGA are that it requires a high vacuum and is usually less sensitive than a helium leak detector.

In use, the RGA spectrum is examined to see if a significant leak exists. Usually the most plentiful gas in a vacuum vessel is water vapor, from outgassing of surfaces inside it. If the predominant gas is air, indicated by nitrogen and oxygen peaks, there is surely an air leak. Sometimes the oxygen will be low or absent because it is reacting with heated material in the furnace. If most of the gas is a backfill gas such as argon, there is a leak in the gas shut-off valve. Water leaks are harder to detect but sometimes changing the water pressure will result in a change of water vapor level.

Having determined that a leak exists, the second step is to locate it. Air leaks can be located using a tracer gas, just as with the helium leak detector, which is applied to each exterior area where a leak might exist. The RGA is operated to monitor the mass where the tracer gas peak is found. When the tracer gas enters the leak, the RGA output will increase. The tracer gas can be any convenient gas except the leaking gas itself.

A second method is to blanket each suspected leak area with a gas, while monitoring the level of leaking gas. When the leaking gas signal disappears, the leak has been located. For example, an air leak can be detected by flooding suspected areas with nitrogen or argon, while watching the oxygen peak on the RGA. When it disappears, the leak is in the area being flooded.

If the leak detection sensitivity must be quantified to prove that all leaks greater than a given size can be detected, a calibrated helium (or other tracer gas) leak should be installed on the vacuum vessel for comparison.

Partial-Pressure Measurement. Partial pressure refers to the concept that in a mixture of gases each species has a partial pressure proportional to its concentration. The sum of these partial pressures is equal to the total pressure of the gas mixture. The measurement of true partial pressures with an RGA is relatively difficult. Some instruments apply a built-in scaling factor to all ion currents and label the results in pressure units. Others provide for entry of sensitivity and gas species constants so that true partial pressures can be displayed.

The factors affecting peak height are:

1) Partial pressure of a given gas species
2) Ionization properties of the gas
3) Ion transmission of the mass filter, a strong function of ion mass; and
4) Sensitivity and linearity of the sensor

Sampling and Pressure Conversion. Vacuum furnace operations that take place at pressures too high (greater than 5×10^{-5} or 1×10^{-4} torr) for direct measurement with a quadrupole RGA can be monitored if a sampling system or pressure converter is used. These devices consist of a small pumping system to maintain a vacuum for the RGA sensor, with a controlled leak to admit a continuous sample of furnace atmosphere into the apparatus. In a properly designed sampling system or pressure converter, the partial pressure of each gas species at the RGA is proportional to its partial pressure in the vacuum furnace but is reduced to a level where the RGA can properly measure it. The pressure reduction ratio is equal to the ratio of leak size to pumping speed. There are two cases: (1) a single-stage system is used for furnace pressures in the range 1×10^{-4} torr to 10 torr, and (2) a two-stage system for pressures from 1 torr to 1 atm.

A single-stage system has a simple orifice for controlling gas input to the RGA pumping system, sometimes called a pressure converter. The inlet pressure is limited to approximately 10 torr by practical orifice sizes and the need to maintain molecular flow. If molecular flow is not maintained, the gas composition is shifted toward gases of high molecular weight. Commercial devices may incorporate several orifices for different pressure levels, with valves to select the desired orifice. Usually a large valve for high-vacuum operation is included. A pressure converter is normally coupled directly to a port on the vacuum vessel or furnace.

A two-stage pressure reduction system has a high-vacuum pumping system to provide a vacuum for the RGA sensor, and second, low-vacuum system to lower the sample gas pressure to a level suitable for a molecular flow leak. The first stage of pressure reduction is carried out in viscous flow so that sample components are not separated. The inlet leak can be an orifice but often is a capillary tube permitting the sample to be drawn from a moderate distance, such as 10 ft or so. The size of the inlet tube orifice is determined by the inlet pressure. Commercial devices offer one or more inlet tubes controlled by valves, sometimes with automatic pressure regulation. Combined single-stage and two-stage systems are also available. Two-stage systems are usually mounted on a portable cart.

Mobile RGA Station

Users often request a mobile RGA station that can be moved to a vacuum system needing service and connected with a flexible hose to make diagnostic measurements of the vacuum environment. Despite the obvious utility of such a device, its performance is severely limited. At pressures above 1×10^{-4}, the inlet of gas to the quadrupole must be reduced by a suitable leak so that its pumping system can keep the sensor at an acceptable vacuum level. The small gas flow of the leak means that gas species must travel through the sample line by diffusion. The sampling system takes a long time to pump out the volume of gas contained in the sample hose. Diffusion works fairly well at pressures below 1×10^{-1} torr, but in viscous flow, diffusion through the length of the sample hose is a very slow process. At high vacuum, where pressure reduction is not needed, the outgassing of the sample line tends to mask the common gases from the furnace volume almost completely. Its use is limited to detecting those gases that are not typical outgassing components.

• 8 •

Temperature Control Systems

James M. Sullivan, Honeywell Inc.

Introduction to Furnace Control

Control of a vacuum furnace requires the control of two variables, temperature and atmosphere, in relation to time. In a vacuum furnace these factors are interrelated in its operating sequences. Temperature control interrelates with vacuum level to delay application of power to the heating elements until the proper vacuum level is attained. Temperature and furnace heating are controlled over time to minimize thermal stress in the workload and often to allow sufficient time for the workload to reach thermal equilibrium, finish outgassing, burn off binder material, etc. If sufficient outgassing occurs it may be necessary to interrupt heating until the required vacuum level is reestablished. Figure 8.1 shows the temperature-vacuum-time relationship.

In batch vacuum furnaces temperature is frequently programmed over time with a temperature set point programmer. Continuous vacuum furnaces rely on the timed rate of workpiece movement through the various furnace heating and cooling zones. Temperature control is just one element in the control of vacuum furnaces.

Temperature Control Loop

The basic temperature control loop consists of a primary sensor, a temperature controller and a final control element (Fig 8.2). The primary sensor in vacuum furnace applications is a thermocouple which provides a dc millivolt signal representative of the furnace temperature to the temperature controller. This instrument compares the furnace thermocouple temperature signal with the temperature set point value of the controller and provides a control signal to the final control element. The controller output signal is

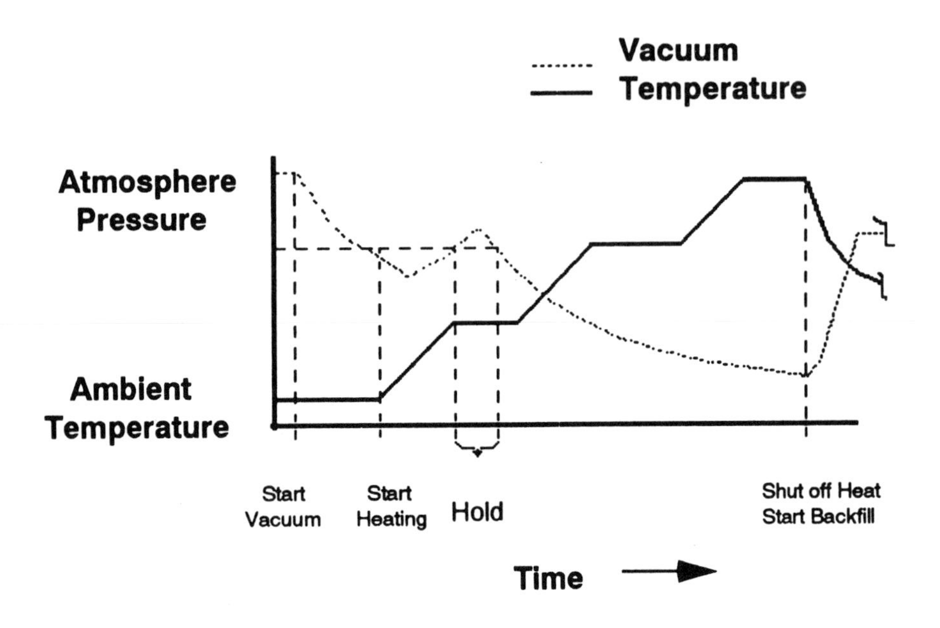

Fig 8.1 Temperature-vacuum-time relationship in a typical process cycle

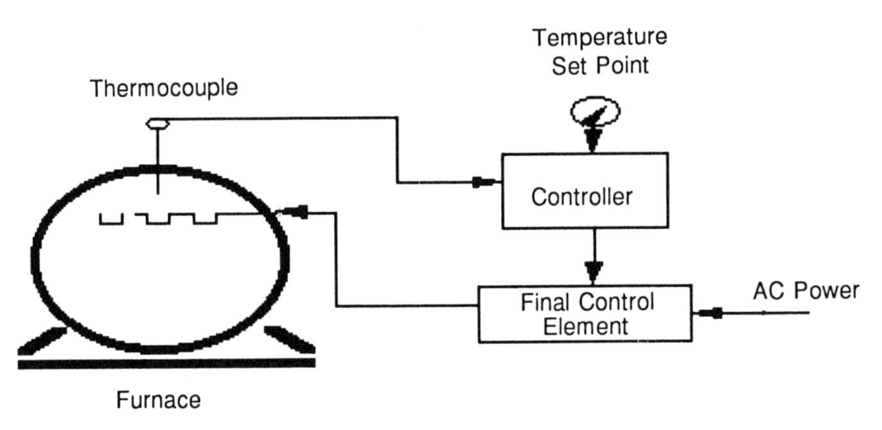

Fig 8.2 Basic temperature control loop

representative of the difference between the desired set-point value and the measured thermocouple signal. The final control element regulates the power supplied to the heating element; its output is proportional to the magnitude of the controller output signal.

Control Loop with Auxiliary Devices. Auxiliary devices used with the basic instruments include temperature recorders, set-point programmers and high limit controllers (Fig 8.3). The recorder may be a single pen, double pen, or multipoint instrument. Selection depends on the application and the process documentation it requires. Control temperature, vacuum level, and workload temperature can be recorded. Figure 8.3 shows a single thermocouple for control and one for the high-limit controller; however it is common practice to employ multiple thermocouples for heat zone control and to scan the zone for minimum, maximum, and average temperature and for zone uniformity. In many applications a Type S (Platinum-Platinum, Rhodium) thermocouple is applied for control and Type K (Chromel-Alumel) thermocouple for workload monitoring.

Set-point programmers vary the temperature vs time and provide storage of programs for operator recall, verification and operation. Set-point programmers send a set-point drive signal to the controller. A control program consists of Ramps (ramp rate or ramp time) and Soaks (soak temperature and soak times). Figure 8.4 is an example of a 9-segment temperature vs time program with event switches. Event switches are digital output signals which can be configured as part of the program to actuate on time,

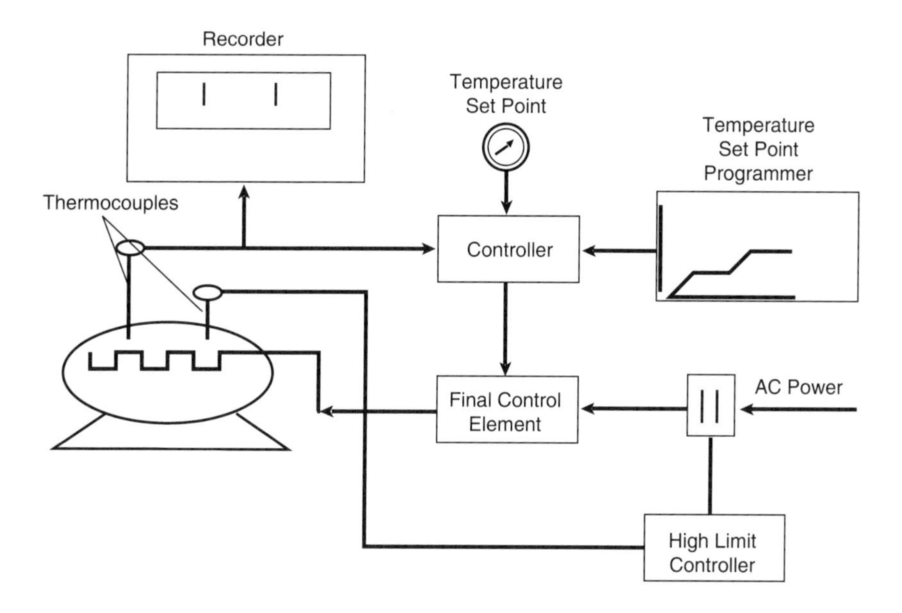

Fig 8.3 Control loop with auxiliary devices

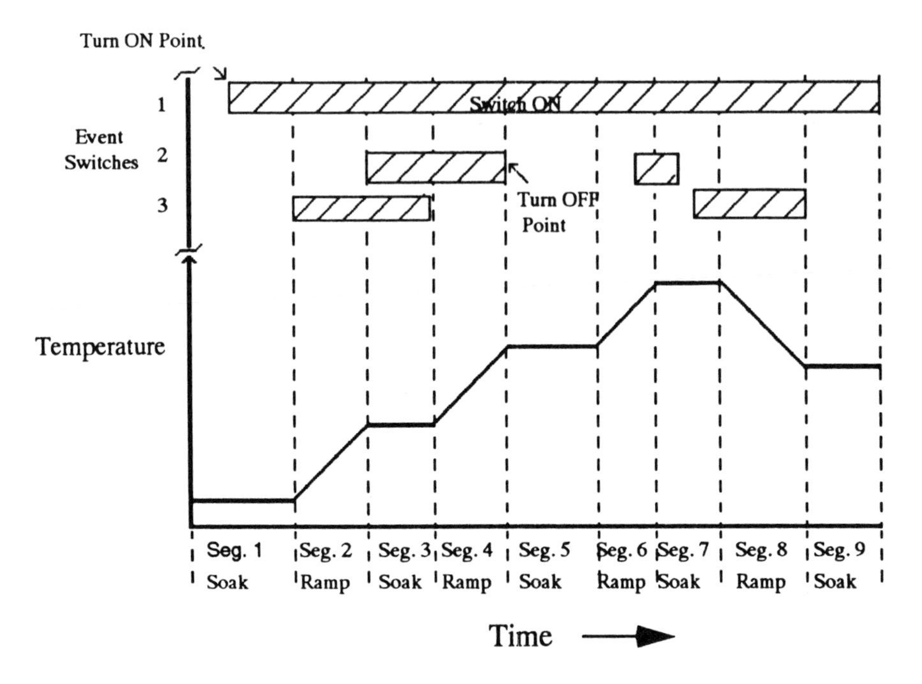

Fig 8.4 Nine-segment temperature vs time program with event switches

temperature value or temperature deviation from the set-point value. Event switches are employed to sequence other furnace operations to the temperature program.

The Batch Vacuum Furnace Control System

In batch vacuum operations, Fig 8.5, the temperature controller is frequently combined with a set-point temperature programmer in a single instrument. The temperature programmer-controller of preference is a digital instrument with a high degree of measurement and control accuracy, usually within 0.56 °C (1 °F) to the reference thermocouple curve. This high accuracy allows the furnace to be programmed over its full operating range while maintaining the relationship between set point temperature and zone temperature uniformity at all operating levels. Additional programmer-controller features include the ability to create and store multiple programs for simple push button recall, continuous display of the program status, and front panel or remote control of operation modes, i.e. Run, Hold, and Reset.

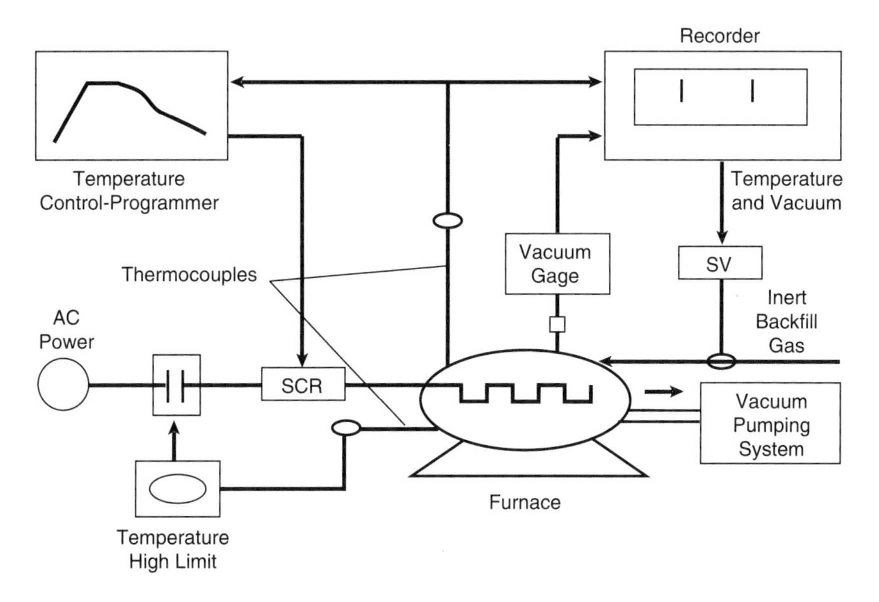

Fig 8.5 Vacuum furnace control system

Temperature Measurement and Heat Transfer

In a vacuum, heat transfer by conduction or convection is practically non-existent. Radiation is the only method of transferring heat from the heating elements to the workload or the temperature sensor. At low temperatures, especially below 480 °C (900 °F), the response to an increase in the heating element temperature will not have an immediate effect on the workload temperature. Conversely, at high temperatures the response is very fast. Furnace temperature control systems are designed to (1) prevent overheating of the workload and furnace heating elements on start-up and (2) obtain optimal response at all temperature levels. Standard operating practice requires the thermocouples to be physically positioned as close as possible to the work to accurately reflect its temperature.

Primary Temperature Sensors

As in the other types of furnaces the thermocouple is the primary sensor used in vacuum furnaces. Although radiation pyrometers are sometimes used, their external mounting requirements mandate sighting on the load through a window. This can present problems due to moisture condensation on the window. In addition, internal gases and water vapor absorb radiated energy and can cause inaccurate measurement.

Thermocouples

A thermocouple is composed of two dissimilar metal wires welded together in a circuit, as shown in Fig 8.6. Whenever a difference exists between the measuring junction and the reference junction, the circuit develops a small dc voltage, representative of temperature at the measuring junction.

Furnace Control. The vacuum furnace control thermocouples are generally made of platinum or platinum and rhodium (type R or S), with a recommended temperature range of –20 to 1480 °C (0 to 2700 °F), and are connected to the temperature controller and the high-limit controller. Thermocouple selection is based on the required accuracy, cost, operating range, upper temperature limit, atmosphere (reducing, vacuum, oxidizing), and mounting method. Platinum control thermocouples are sheathed in ceramic protecting tubes. In all applications the temperature and vapor pressure must be considered in selecting the thermocouple type and protecting tube.

Workload. Workload thermocouples are generally type K (Chromel-Alumel) with a recommended temperature range of 20 to 1095 °C (0 to 2000 °F)

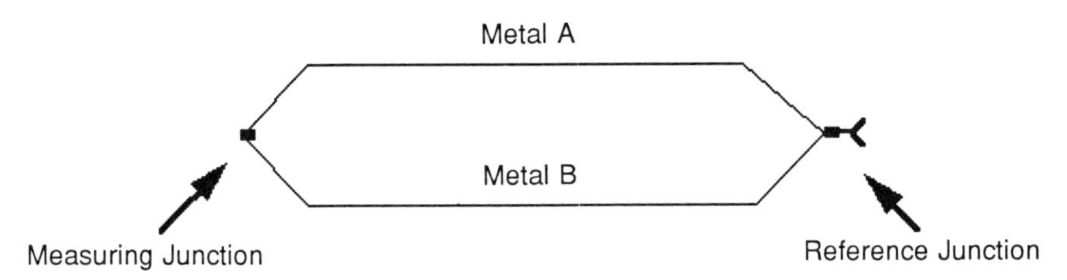

Fig 8.6 Thermocouple circuit

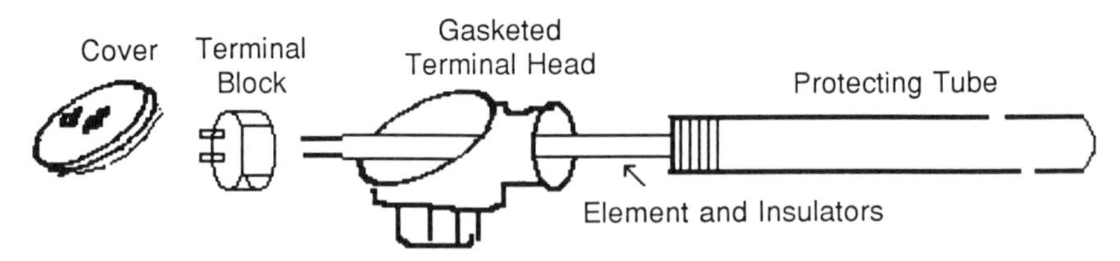

Fig 8.7 Thermocouple with protective tube

for #14 gage wire. Chromel-Alumel is selected due to its relatively low cost (10 to 20% of platinum) and its applicable temperature range. Most applications use multiple (2 to 12) thermocouples attached directly to various parts of the load, with their outputs recorded on a multipoint temperature recorder, or utilized for furnace control, or both. Workload thermocouples take various forms, including bare-wire ceramic-bead insulation, high-temperature refractory-cloth insulation, and metal-sheathed types. Metal-sheathed thermocouples (Fig 8.7) are used in applications where the workload material might contaminate or otherwise destroy a bare-wire type. Thermocouple wire sizes are #8, #14, and #20 gage. The larger the gage size, the higher the recommended application temperature. As an example, #8 gage type K can be used up to 1260 °C (2300 °F), but #20 gage is limited to 980 °C (1800 °F). The larger the wire size the slower the speed of response to temperature changes. Workload thermocouples have a limited service life dependent on the temperature operating level, thermocouple construction, and vacuum level.

Extension Wire. Thermocouple extension wire has electrical characteristics similar to those of the thermocouple. Its purpose is to extend the thermocouple reference junction from the connecting head to the measuring instrument reference junction compensator, located at the input terminals of the instruments (Fig 8.8). The measured temperature is the voltage difference between the thermocouple measuring junction and the reference junction in the instrument. The extension wire is usually furnished in the form of a matched pair of conductors with insulation that meets the service needs

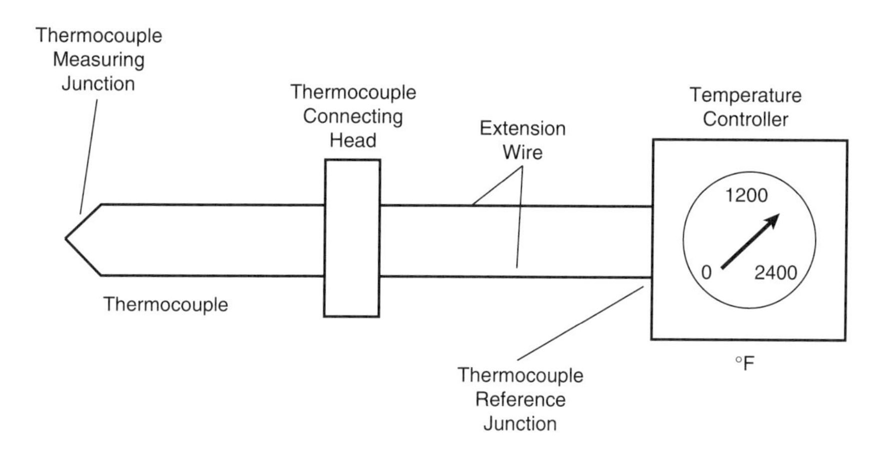

Fig 8.8 Single thermocouple circuit

of the application. Recommended practice is that the temperature at the thermocouple extension head and the thermocouple extension wire should not exceed 205 °C (400 °F). Extension wire should be of the same type and calibration as the thermocouple; copper extension wire should not be used.

Radiation Temperature Detectors

There are two principal types of radiation detectors: (1) the optical infrared pyrometer, and (2) the radiation pyrometer. Radiation pyrometers are capable of measuring a wide temperature range and are suitable for control.

Optical Pyrometers. Optical pyrometers are narrow-range detectors and their application is primarily for measurement and test, rather than control. Optical radiation devices are applied when the temperature to be measured exceeds the upper limit for thermocouples, and the speed of the work is too fast, or the thermocouple cannot be physically located close enough to the load to accurately reflect its value.

Radiation Pyrometers. Hot objects generate radiant energy in the infrared, visible, and ultraviolet spectra. These waves radiating through space will heat any other objects they strike. Radiated energy is the primary heat transfer method in a vacuum, and this is the way the furnace heating elements heat the workload, and the way the radiation pyrometer senses the temperature of the work. It has a lens that focuses radiated energy on a thermopile (Fig 8.9), an assembly of thermocouples connected in series, to absorb the energy, strengthen the signal, and generate a dc millivoltage representative of the temperature value. The output signal of the detector is proportional to temperature as a fourth power function and is non-linear.

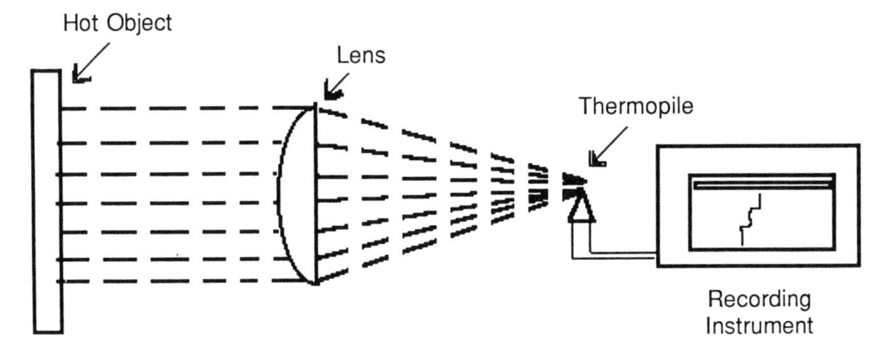

Fig 8.9 Radiation pyrometer operating principle

The sensor signal output can be increased by increasing the number of thermopile detectors, but increased mass reduces the response speed of the unit to temperature changes. Thin-film technology used in recent designs has minimized response time and signal strength limitations.

Three-Mode Temperature Controllers

The temperature controller applied to vacuum is generally a three-mode controller (proportional, reset, and rate) with a current ouput control signal (Fig 8.10). A key feature of microprocessor-based controllers is process auto tuning capabilities. Many auto tune strategies exist. The key element of this feature is a software program that measures the response of the process to upsets and applies the required controller tuning parameters. If auto tune features do not exist in the controller, or manual tuning is employed, the following discussion on controller tuning may be helpful.

Adjustment of the controller mode settings (controller tuning) determines how the furnace temperature will respond to operating conditions. These include (1) set-point changes, and (2) furnace loading and load size.

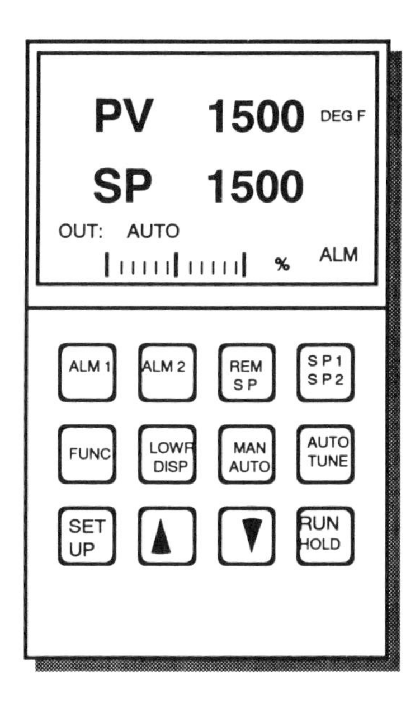

High Measurement Accuracy

Dual Set Point & Temperature Displays

Auto Manual Station

Multi. configurable Alarms

Multiple Tuning Constants

Automatic Tuning
Set Point Programming

High Noise Immunity

English Language Prompts

Fig 8.10 Digital temperature controller features

Response characteristics are in terms of the time it takes to achieve and maintain steady control at the desired set point value.

Proportional Band. Proportional band is the reciprocal of gain. Gain is simply the output signal divided by the input. Proportional band is 100 divided by the gain, per the example in Fig 8.11.

Automatic Reset (also known as "Integral Action," i.e., the integral of a step is a ramp). Automatic reset will sense if an error or offset exists between the measured temperature and the temperature controller set point value. Reset action will change the power settings to the heaters until the desired temperature is obtained. Controllers with automatic reset will reposition the heater power setting at a speed proportional to the size of the error that exists. No error, no reset action.

Proportional Plus Reset Control. The proportional plus reset modes are illustrated in Fig 8.12. In (a) an error is caused by increasing the set point value at a point in time. The error created is equal to the set point value minus the measured temperature. In (b) the response to the created error is shown under three different values of reset: 0; a mid-level value; and a high value. The proportional response contribution, a step change in heat output,

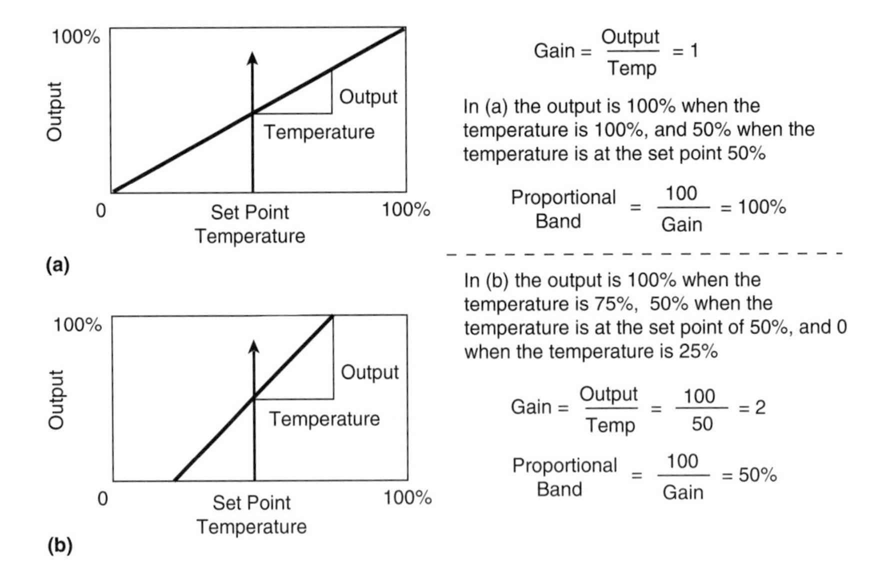

Fig 8.11 Controller tuning

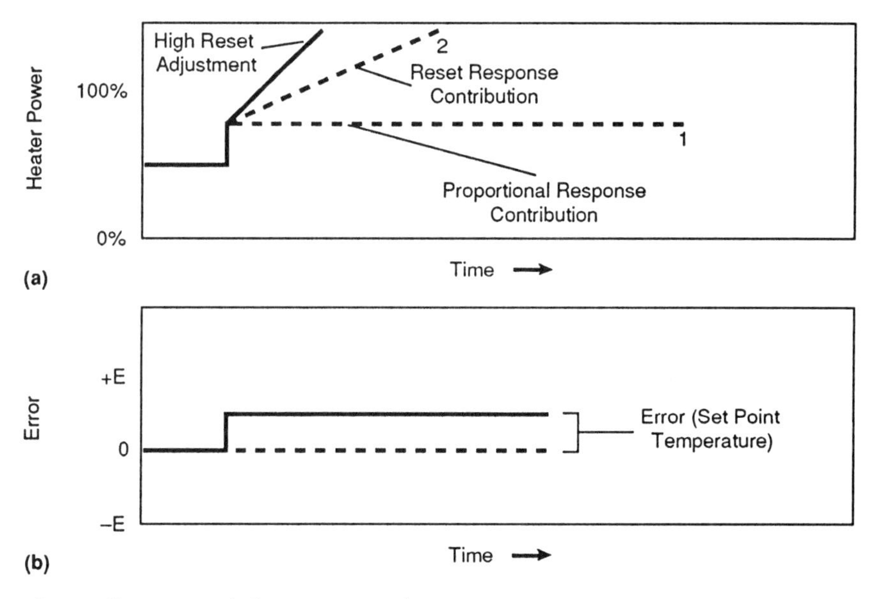

Fig 8.12 Proportional plus reset control action

is equal to the value of controller gain only; the reset contribution is zero. The broken line in the response graph, labeled 2 in Fig 8.12(a), shows the contribution of some intermediate value of reset-plus-gain: a step change plus a ramp in heater output. The solid line illustrates the heater output in response to a high value of controller reset. Adjustments in reset time determine the slope of the ramp, as illustrated by the ramp pictured. Reset time on a controller can be expressed in repeats per minute, or the reciprocal minutes per repeat. A typical 3-mode controller will have a Reset adjustment range of .08 to 10 min.

Recapping the action of the controller to (1) mode settings and (2) the magnitude of the error signal: The response, Fig 8.12(a), is (1) for Gain, a step response in output equal to the set change in set point value, and (2) the reset contribution to a step change is a ramp, Fig 8.12(a). The slope of the ramp is determined by the controller reset mode setting. The combined effect of a controller with gain plus reset is a step change in output signal, plus a ramp of the output equal to the reset value and the magnitude of the error which remains between set point value and the measured temperature.

Rate Action. Rate action, also called derivative, responds to the rate of change in set-point or error signal, by changing the power applied to the

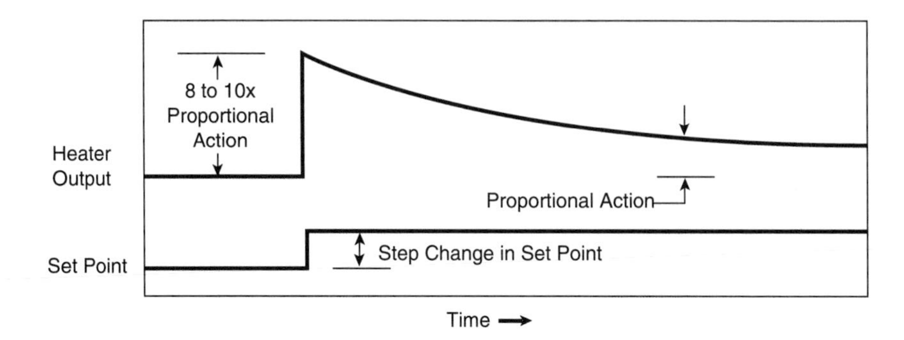

Fig 8.13 Rate action response to a step change

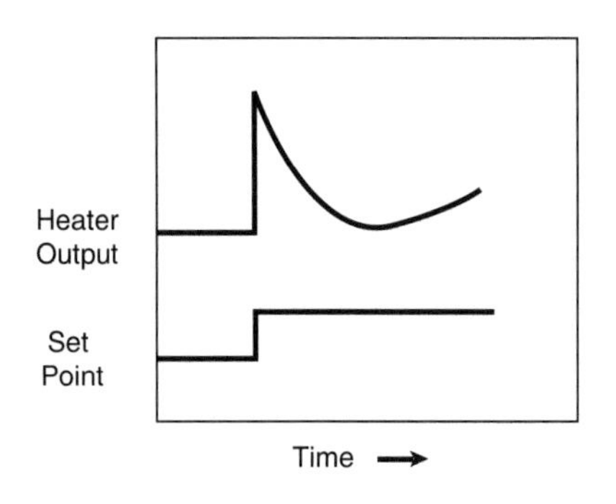

Fig 8.14 Three-mode control action

heaters at a speed proportional to the change. The most notable occurrences are in batch vacuum furnaces where the set point is frequently changed by the operator or time-temperature programmer. Rate action responds to rapid changes in error, with a rapid change in heater output to, in effect, over-correct for the change. Rate action, therefore, over-corrects, and is then removed, leaving proportional band or proportional plus reset action to bring the process back to set point. This action is illustrated in Fig 8.13.

The heater output with rate action exceeds that which would have occurred with proportional action alone. Rate time is the length of time for the decay to reach the position it would have assumed with proportional action alone. Rate time adjustment range is typically .08 to 10 min.

Proportional Plus Reset Plus Rate. Figure 8.14 shows the action of a three-mode controller in response to a step change increase in the set point position. The heater output is determined by adding the effects of the three modes.

Controller Automatic Tuning. There are many names given to the automatic tuning feature of digital controllers. Automatic controller tuning relieves the furnace operator and the process engineers of the burden of determining the proper settings for controller Gain, Rate and Reset values.

The term "automatic tuning" generally describes the continuous or period adjustment of controller settings in response to process conditions. During furnace start-up, after set point changes, and during process disturbances the controller analyzes the process and outputs a controlled response.

Final Control Elements

Vacuum furnaces use 3-phase low-voltage power to the heating elements. Proportional control of the power is achieved by use of variable reactance transformers, saturable core reactors, and silicon controlled rectifiers. All of these devices can proportionally control large amounts of ac power in response to the output signal from the temperature controller.

Saturable Core Reactors

A saturable core reactor is a device that can vary the electric power to the furnace heaters in a manner similar to an adjustable valve in a water pipe. A saturable reactor is a transformer type device in which the impedance is varied by changing the magnetic characteristic of the core with dc voltage. Its method of delivering proportioned power to the load is to let a small control signal affect the degree of saturation of the magnetic core, which in turn affects the ability to pass larger currents to the load.

A block diagram of a saturable reactor control system (Fig 8.15) shows a magnetic amplifier between the temperature controller and the saturable reactor. The magnetic amplifier is used to provide a variable voltage, 0 to 75 vdc, to drive the reactor. In the illustration the load requires an automatic current limiter. This requirement exists when the load is a heating element with resistance which changes radically as its temperature increases from ambient to normal operating temperatures. A change of 2 to 1 is not considered radical, but heaters made of molybdenum, platinum, tungsten, and other metals can change by 10 to 1 or more. Thus, when the heating elements are cold, they can draw as much as ten times their rated current and overload the reactor if current limiting were not used. Current limiting also prevents excessive thermal shock to the heating elements, which would shorten their life.

Variable Reactance Transformers

Saturable reactors are commonly coupled with the primary windings of stepdown transformers to obtain the low voltages necessary for the vacuum heating elements. The presence of a transformer primary as the reactor load is treated as the final load itself. A variable reactance transformer (VRT) is similar to a saturable reactor in series with the primary of a transformer.

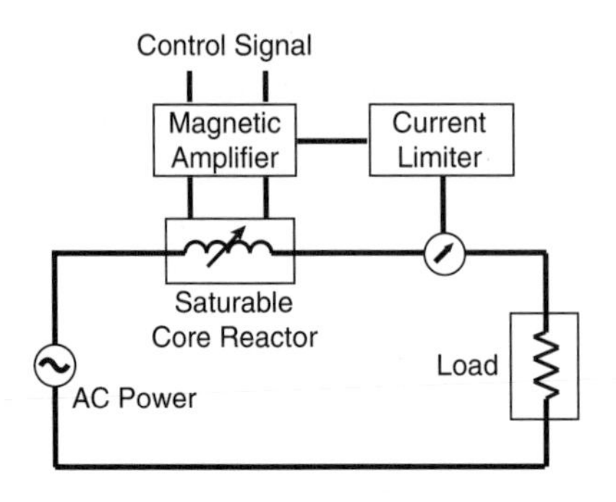

Fig 8.15 Saturable reactor control system

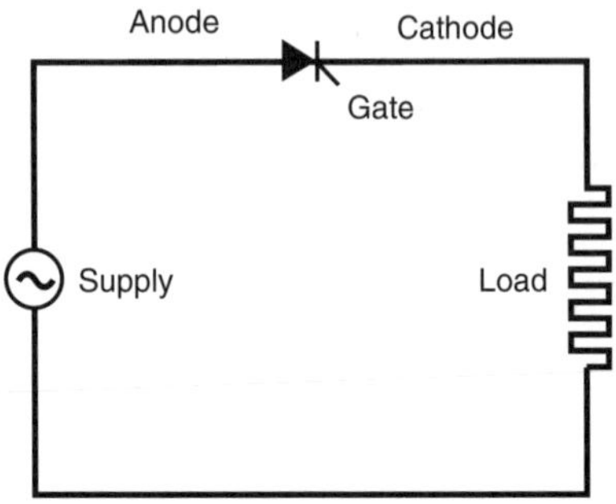

Fig 8.16 SCR circuit

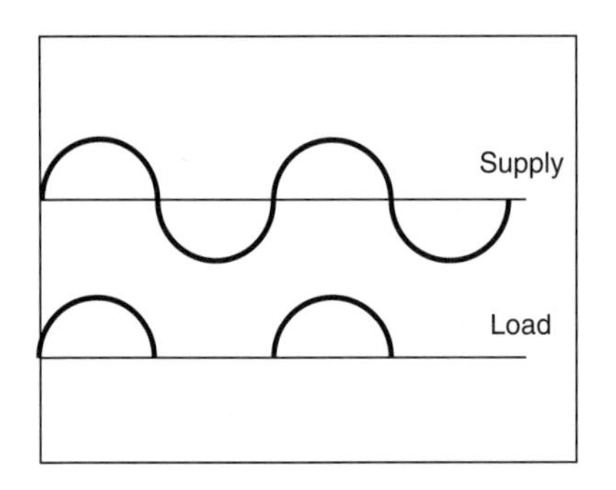

Fig 8.17 SCR supply vs load waveforms

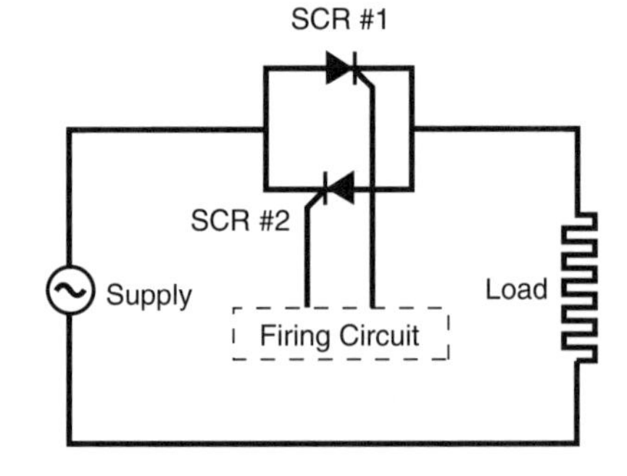

Fig 8.18 SCR firing circuit

Basically, it is a transformer which can be designed over a wide range of ratios and has an infinitely variable coupling between the primary and the secondary. It makes a greater or lesser amount of current available to the secondary, depending on the dc control signal applied.

Silicon Controlled Rectifiers

A silicon controlled rectifier (SCR), shown in Fig 8.16, is basically a solid state switch capable of passing current in one direction only. As with a diode, the anode must be positive with respect to the cathode for conduction to begin. The SCR, however, has another dimension in that the gate signal controls the turn-on point. Proper polarity must be observed and a gate signal must be present for conduction to occur. Once the SCR has fired, the gate signal may be removed. It will continue to conduct until current flow decreases below a minimum required to sustain it. As the current waveform decreases toward zero, the SCR will turn off (Fig 8.17). A gate signal and proper polarity with respect to anode and cathode is required for the SCR to turn on.

Two SCR devices in an inverse parallel connection (Fig 8.18) will produce full-save control. Each SCR conducts only on that half-cycle where proper polarity is applied to its anode and cathode. Controlling the output of a pair of SCRs in this type of circuit is accomplished by varying the timing of the gate pulses applied to both of them, as in Fig 8.19. The level of output depends on when the gate pulse is applied.

There are two methods of controlling the output from an SCR, both associated with the timing of the gate or firing voltage, known as zero crossover or burst fire control and phase angle firing. Phase angle firing is employed in vacuum furnace applications due to the requirement for current limiting when the cold heater elements are first energized. Figure 8.20 shows phase angle firing. At T_1 and T_2 the gate pulse occurs at the beginning of each half cycle of the wave. At T_3 and T_4 the gate is delayed until the half wave is at its peak or 50% point. Comparing the gate signal to the load shows that the load can be controlled by controlling the timing of the gate from full conduction to full off.

SCR applications which require a soft start or current limiting must use phase angle firing as opposed to zero crossover firing.

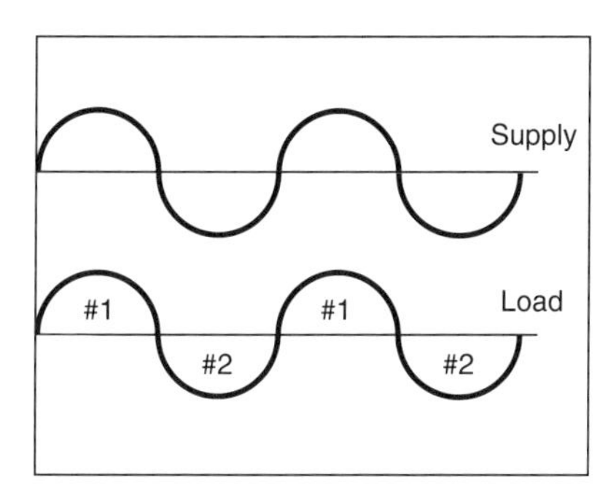

Fig 8.19 SCR firing circuit supply vs load waveforms

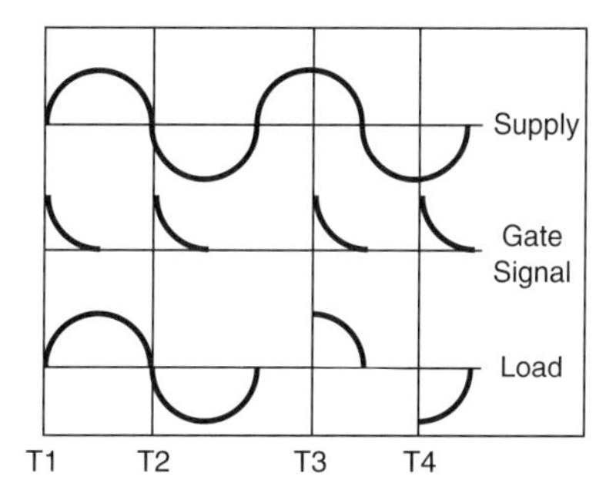

Fig 8.20 SCR phase-angle firing waveforms

Operating Characteristics

Table 8.1 compares the operating characteristics of saturable reactors and VRTs vs SCRs.

Integrated Furnace Control Systems

The integrated control systems available from all major instrument vendors combine LOOP control (multiple loops for temperature, vacuum, load profiles, etc.); LOGIC control for machine sequencing, control and safety interlocking; and ADVANCED OPERATOR controls and process/machine interfacing.

Microprocessor-based control systems are available which offer furnace designers and users the option of combining the LOOP, LOGIC and ADVANCED OPERATOR interface devices in a single cost-effective package. Major benefits are (1) A single master controller can be programmed to perform all furnace operations, (2) It provides critical information to the operator and prevents unsafe operation, and (3) It initiates safety shutdowns in the event of equipment malfunction. In addition, all process and sensor based information is readily accessible by supervisory or management level computers through standard communication ports.

Table 8.1 Operating characteristics, saturable reactor or VRT vs SCR

Saturable reactor or VRT	SCR
Requires an intermediate amplifier, such as a magnetic amplifier.	Requires no intermediate magnetic amplifier. Driven by the 4-20 ma controller output signal.
A non-linear device. Not a tight shutoff device.	A linear device.
Must be matched to the load.	Load matching not required.
Response time: .5 to 6 seconds.	Response time less than 1 second.
Large and heavy.	Requires added circuitry for the current limiting option.
Some inherent current limiting ability.	Generates higher level of electrical noise.

Temperature Limit Controller

Applied as a high-temperature limit controller to interrupt power to the heaters in the event furnace temperature exceeds a preset value, this device provides a latching relay which is activated whenever the temperature process variable goes above or below a preset set point value. An alarm indicator lights when the output is activated. Manual reset is accomplished via a button or key on the front of the instrument, or by an external contact.

• 9 •

Commercial Practice

Wilbur T. Hooven III, Hooven Metal Treating, Inc.
Jeremy St. Pierre, Hayes Heat Treating Corp.
Howard Sanderow, Management and Engineering Technologies

Common Metal Systems

Given the proper vacuum furnace and ancillary equipment, just how is a vacuum furnace cycle run on common materials used in industry today? The intent of this chapter is to answer that question by reviewing how common vacuum heat treating, brazing, and sintering cycles are performed on materials of general interest. This chapter is divided into four sections, covering heat treating of oil-quenched and gas-quenched metals, brazing, and sintering. It is assumed the reader has a reasonable understanding of the fundamentals of vacuum technology and how it is applied to heat treating, brazing, and sintering. While there is a wide variety of types of vacuum furnaces on the market and in general use, the most commonly used type of vacuum furnace has been assumed for each of the processes discussed.

In every case, for each of the processes discussed in this chapter, due regard must be given to the careful placement of the workpieces on the hearth of the vacuum furnace. It must be remembered that heating in a vacuum furnace depends primarily on radiation for uniform heating, with only a minor contribution from part-to-part or part-to-fixture conduction. Consideration must be given to the thermal gradients of both the workload and the furnace during heating and cooling. Ideally, the workpieces will be sufficiently spaced to permit the most uniform heating and cooling of them. However, this usually unfavorably impacts the economics of the process, and thus a compromise must be made between ideal loading and economics. Failure to address these two important considerations when determining furnace loading, can cause the results of the vacuum process to be less than satisfactory, from either a quality and/or an economic standpoint.

For repeated production cycles, it is usually adequate to establish by use of multiple load thermocouples, together with process capability evaluation, an optimum load distribution, weight, and cycle time relative to the control thermocouple. Identical workload cycles can then be run by control of the furnace thermocouple alone. With any furnace that involves movement of the workload (e.g. an oil-quench furnace) control by the furnace thermocouple alone is usually the only method possible. Whenever identical furnace loads are not the normal process and/or it is possible to safely instrument the load with thermocouples, it is prudent to rely on individual load thermocouples (two as a minimum, preferably four) to assure proper surveillance of the thermal cycle.

Heat Treating Oil-Quenched Metals

Metals with mechanical properties that can be obtained only by oil quenching form the largest group of heat-treatable metals on a tonnage-processed basis. These include low- and medium-alloy steels, tool steels including high-speed steels, corrosion/heat resistant steels and alloys, and precipitation-hardening nonferrous alloys. Oil-quenching these materials has been the basis for the majority of heat treating history. Typically oil-quenched metals are processed in air, salt, or most commonly, in a protective atmosphere. Although they do not generally require processing in a vacuum furnace and for the most part can be processed at a lower cost in an air, salt, or atmosphere furnace, there is an increasing trend to using vacuum equipment for them to produce a superior product when the cost can be justified.

It should be noted that it is possible in many cases to process metals that are traditionally oil quenched in a vacuum furnace with a gas quenching system. The hardenability of the material being processed, the section size of the workpieces, the size and configuration of the workload, and the performance capability of the furnace to be used must all be carefully considered when deciding on the quenching method to be used.

Low- and Medium-Alloy Steels

It is possible to oil-quench virtually any low- or medium-alloy steel and obtain full mechanical properties in a vacuum furnace with oil quenching capability. This is, of course, provided the material has sufficient hardenability for the section size being processed with the quench rate available. It is even possible to satisfactorily harden many water-quenching materials in such a furnace.

Oil-quench vacuum furnaces with graphite heating elements and insulation are the most commonly used type of furnace for oil-quenching low- and medium-alloy steels. These furnaces can have the heating chamber separated from the quench chamber by only a heat resistant door or, in addition, a vacuum-tight inner door. While a vacuum-tight inner door will certainly improve productivity, as the heating chamber does not have to be cooled to room temperature each cycle, the inner door does add significant expense to the furnace. See Figs 9.1 and 9.2 for cross sections of typical oil-quench vacuum furnaces with and without a vacuum-tight inner door.

Oil-quench vacuum furnaces can produce satisfactory work with only a mechanical pump, which can produce a working vacuum of approximately 300 to 500 microns. The addition of a blower, which will increase the vacuum level to less than 50 microns, while desirable, is not essential. A diffusion

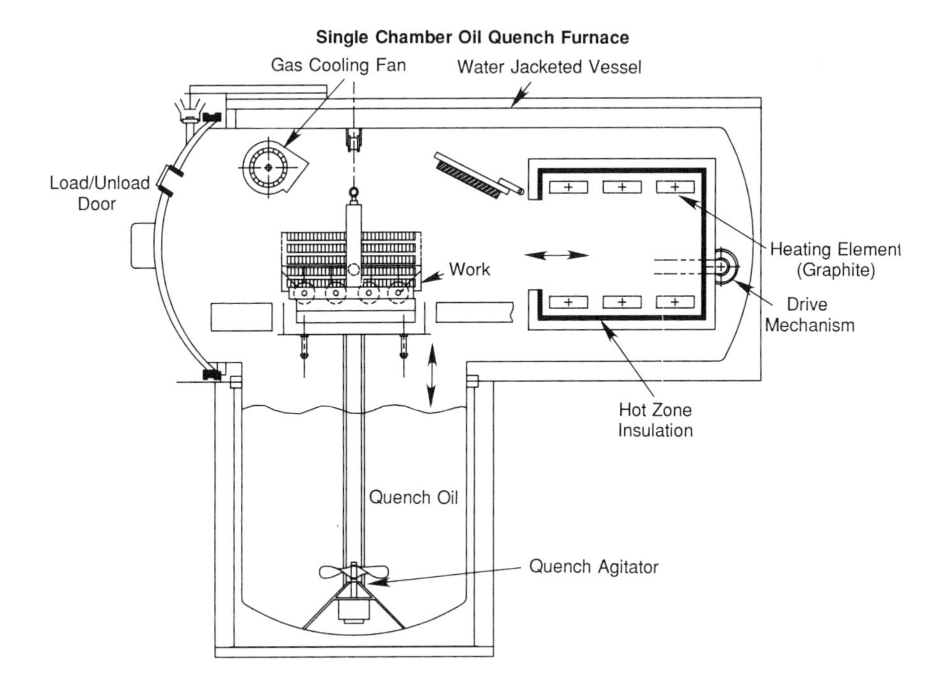

Fig 9.1 Section of oil quench vacuum furnace with vacuum-tight inner door

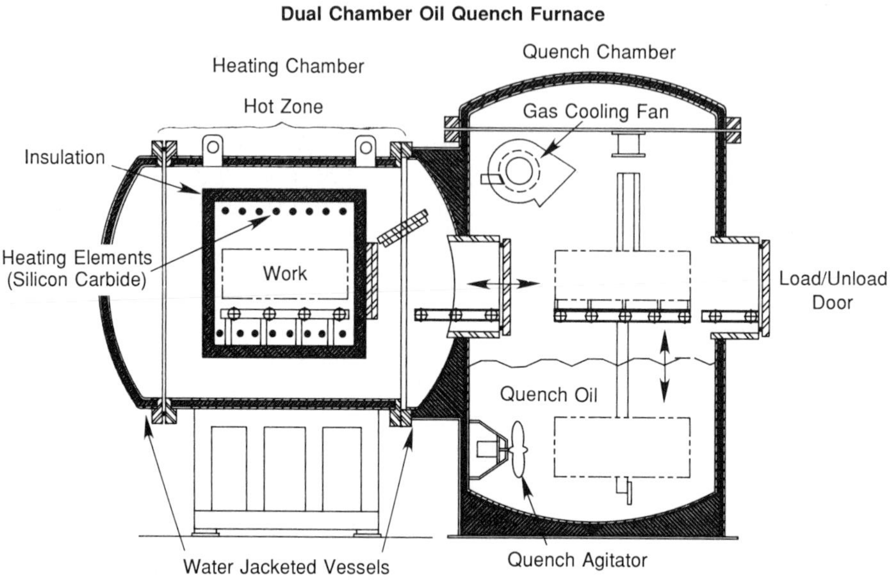

Fig 9.2 Section of oil quench vacuum furnace without vacuum-tight inner door

pump is not required, or even desirable, for processing low- and medium-alloy steels. A controlled partial pressure of 50 to 500 microns is often useful for dealing with workloads that are outgassing excessively. A controlled partial pressure is not required to prevent vaporization, as vaporization is extremely unlikely at the vacuum levels and temperatures normally used to process low- and medium-alloy steels.

Normally a medium-fast quench oil is used for quenching these materials. The ability to keep the oil in the 40 to 55 °C (100 to 130 °F) range is desirable. A partial pressure of 20 in. Hg of nitrogen is normally maintained over the oil during quenching. Argon is occasionally used, but it is expensive and poses a potential safety hazard because it is heavier than air. The gas used to backfill the furnace should be from a cryogenic or other high-purity source. It is possible to manipulate, to a limited degree, the quenching characteristics of an oil by varying the pressure over it.

Carburizing, carbonitriding, and ferritic nitrocarburizing, processes commonly performed in atmosphere furnaces, can also be performed in suitably equipped vacuum furnaces. The process involves heating the workpieces to be carburized or carbonitrided in vacuum, backfilling the heating chamber to a partial pressure with a carbon or carbon/nitrogen containing gas and then quenching as previously described. The process offers excellent uniformity and control but is more costly than when performed in an atmosphere furnace. Thus, it is normally limited to high value-added workpieces such as aircraft quality gearing and bearings.

Tool Steels

Oil-hardening tool steels are hardened in much the same way as low- and medium-alloy steels. The austenitizing temperature range is wider for these materials. Thus, it is important to understand how the heating efficiency of the furnace to be used varies with temperature. Preheating is advisable particularly if the workpiece involves any major section size changes or the loading is such that certain parts of the workpiece or workload will come to temperature first. Normally only a single preheat is required.

The addition of a blower to the vacuum pumping system is desirable to ensure bright surfaces on the finished workpieces. Tool steels processed in an oil-quench vacuum furnace without a blower sometimes have a very light blue/grey color. This is normally of no metallurgical significance. A diffusion pump is not necessary for hardening tool steels. A controlled partial pressure, up to 500 microns, is often useful for dealing with workloads that are outgassing. A controlled partial pressure is not required to prevent vaporization, as vaporization is extremely unlikely at the vacuum levels and temperatures normally used to process oil-hardening tool steels.

It is possible to process air-hardening grades of tool steels in an oil-quench vacuum furnace. The technique used is an interrupted oil-quench. The workload is quenched in the oil for only a few seconds and then removed. The temperature of the workpiece is allowed to equalize and the workload is again lowered into the oil. The procedure is repeated several times until the workpiece is cool enough to be left in the oil, or it is safe to remove the workload from the furnace. Improperly applied, this technique can result in distorted and/or cracked workpieces. Workpieces processed using an interrupted oil quench normally have a light "burnt oil" film on the surface. While rarely of any metallurgical significance, the finish is not attractive.

High-Speed Steels

High-speed steels can be hardened in an oil-quench vacuum furnace, and for some applications this is still the preferred method. However, it is now much more common to harden high-speed steels using a gas-quench vacuum furnace.

High-speed steels are processed in much the same way as tool steels. The austenitizing temperature range is much higher for them. Thus, it is very important to understand how the heating efficiency of the furnace to be used varies with temperature. Preheating is required, normally only a single stage, but two preheat stages are not uncommon.

Because of the high austenitizing temperatures used, materials used to support the workpieces must be carefully considered. Most commonly used materials have very little strength at the temperatures used to austenitize high-speed steels. Therefore, an alloy such as Haynes 230 is recommended. Ceramic beads are often used to separate the workpiece from its support, particularly if graphite is used. There is a very good possibility of having at least "sticking" problems, or worse, incipient melting at the surface, if metallic materials to be used have been carburized. To avoid such problems, surfaces touching the workpiece should be painted with a "stop-off" paint consisting of light aluminum oxide slurry suspended in a binder.

The addition of a blower to the vacuum pumping system is recommended for a vacuum furnace to be used for hardening high-speed steels. High-speed steels hardened in an oil-quench vacuum furnace without a blower sometimes have an increased surface carbon content, particularly if the furnace is of the type without a vacuum-tight inner door and it is not kept clean. A diffusion pump is not desirable for hardening high-speed steels. A controlled partial pressure of 50 to 500 microns is often useful for dealing with workloads which are outgassing and for preventing vaporization of alloying elements with a high vapor pressure, such as chromium.

High-speed steels are often oil-quenched using the interrupted oil quench technique described earlier. Again, this technique improperly applied can result in distorted and/or cracked workpieces. Workpieces processed using an interrupted oil quench will have a heavier "burnt oil" film on the surface because of the higher processing temperatures used.

Corrosion-Resistant/Heat-Resistant Steels and Alloys

The majority of the corrosion-resistant/heat-resistant steels and alloys require a rapid quench following austenitizing or solution annealing, to ensure that maximum properties are obtained. In many cases the steel or alloy manufacturer recommends heating in air followed by a water quench. While this method will produce optimum metallurgical results, it is sometimes impractical for net-shape workpieces. Oil-quench vacuum furnaces are ideally suited for austenitizing or solution annealing net-shape workpieces fabricated from many of these materials, because there is no surface oxidation and the quench rate, while severe enough to ensure optimum metallurgical properties, is not so severe that workpiece distortion is a problem.

The austenitizing or solution annealing temperature range for these materials is often quite high, and it is important to understand how the heating efficiency of the furnace to be used varies with temperature. Preheating is recommended to ensure minimal workpiece distortion and thermal cycle uniformity throughout the workload.

The addition of a blower to the vacuum pumping system is required for a vacuum furnace to be used for processing corrosion-resistant/heat-resistant steels and alloys. While a diffusion pump may be desirable for certain materials from a surface oxidation standpoint, these pumps can require much maintenance, particularly if the furnace is of the type without a vacuum-tight inner door and is not kept clean. A controlled partial pressure of 50 to 500 microns is often used for dealing with workloads which are outgassing and for preventing vaporization of alloying elements with high vapor pressures, such as chromium.

Precipitation-Hardening Nonferrous Alloys

Certain precipitation-hardening nonferrous alloys, such as beryllium copper, and magnetic alloys, such as Cunico, can be satisfactorily solution-annealed using an oil-quench vacuum furnace. The vacuum pumping system required will vary with the alloy being processed. Normally a vacuum system consisting of a mechanical pump or a mechanical pump and blower, is satisfactory. A controlled partial pressure of 50 to 500 microns is often used for dealing with workloads which are outgassing and for preventing vaporization of alloying elements with high vapor pressures, such as copper. The temperatures used are varied, and it is important to understand how the heating efficiency of the furnace to be used varies with temperature. Preheating is recommended, to ensure thermal cycle uniformity and minimal workpiece distortion throughout the workload.

Typical Oil-Quench Vacuum Furnace Operating Cycle

Note: Do not attempt to process an oily and/or wet workload. It cannot be stressed strongly enough that the workload must be clean and dry. An oily and/or wet workload could result in damage to both the workpieces and the furnace. Ideally, the workpieces should be cleaned just prior to processing, particularly where humid conditions prevail. Hot-vapor degreasing or an alkaline-water wash followed by hot-air drying are recommended.

1) Set up the workpieces using appropriate fixturing as required. For low- and medium-alloy steels, standard stacking baskets made from materials such as RA85H, RA330, Inconel 600, Inconel 601, or Inconel 800, are commonly used. Fixturing made from cast heat-resistant al-

loys is also commonly used. For temperatures greater than 925 °C (1700 °F), Haynes 230 Alloy or Inco MA 956 are an excellent choice. Molybdenum is normally used for temperatures greater than 1205 °C (2200 °F). Graphite or ceramic are not normally used for fixturing when oil-quenching. It is important to include the weight of the baskets and fixturing when determining the weight of the workload to be processed, and to observe the maximum weight recommendations of the furnace manufacturer.

2) Charge the workload into the furnace. Where possible, verify that the work transfer mechanism is working properly prior to evacuating the furnace. Having a workload get stuck mid-transfer from the heating zone to the quench tank, or worse having the load upset, will normally ensure that work transfer is verified on all subsequent loads. Wipe down door seal surface and check for O-ring damage.

3) Set the over-temperature control at no more than 28 °C (50 °F) above the maximum temperature to be used.

4) Evacuate the furnace. Often backfilling it with either nitrogen or argon to 15 in. Hg after the vacuum level stabilizes, and then evacuating again, will improve the ultimate vacuum and decrease the time required before the cycle can be started. Normally a minimum vacuum level of 50 to 500 microns is required before the cycle can be started. From this point on, furnace operation is usually automatic.

5) Ramp the workload to the preheat or process temperature. Ramping the furnace to the processing temperature rather than unrestricted heating is recommended to minimize overshoot and the possibility of distortion problems caused by overheating. Typically ramp rates of 8 to 14 °C (15 to 25 °F) per min. are used.

6) Soak the workload at the process temperature as required. It is important to note that the workload will lag the set-point by some margin even if a relatively slow ramp rate is used. Time must be allowed for the center of the workload and centers of the workpieces to come to temperature. Soak times of less than 45 min. are not recommended for workpieces even with relatively thin sections. Instrumenting a test load to determine the soak duration of a particular workload configuration is recommended for critical work.

7) At the end of the soak time the load is quenched, usually automatically by the furnace. The quench sequence normally consists of backfilling the heating chamber to 20 in. Hg with nitrogen while maintaining the set-point temperature, opening the heating chamber door, quickly transferring the workload over the quench tank and then quickly lowering it into the tank. The time interval from the heat-

ing chamber door opening to the workload entering the quench oil should not normally exceed 15 s. Longer times than this can result in problems with under-hardening on the surface of the top basket of the workload. Covering this basket with expanded metal screen is often an effective means of preventing heat loss from its surface.

8) Depending on the furnace design, the heating chamber may or may not be cooled to room temperature while the workload is being quenched. When the heating chamber is cool and/or the workload is cool, the workload can be removed from the quench tank. Allowing it to drain over the quench tank for a few minutes prior to removing it from the furnace will significantly reduce drag-out losses and ease subsequent cleaning. It is important to remember that the furnace is full of a gas which does not support life. Two people should always be present when a furnace is being unloaded, particularly with argon, because it is heavier than air.

Heat Treating Gas-Quenched Metals

The modern aerospace industry and World War II provided the technical impetus to develop vacuum processing as a result of more precise material demands in performance, quality, and integrity. Out of this evolved the modern sophisticated and ever-improving gas-quench vacuum furnace (Fig 9.3). Furnaces were developed that could provide vacuum levels high enough to process even the most demanding materials, with quenching rates that could meet the metallurgical requirements for properties. These furnaces reduced or eliminated the distortion common with liquid quenching, and produced clean, finished workpieces requiring no further metal removal.

In addition to advances made possible in the aerospace industry, gas-quench vacuum furnaces have contributed to the development of many product applications of superior quality in medical use, electronics, and other fields. These products are often produced at a lower total cost as a result of no post-heat-treat finishing. Metals include tool and high-speed steels, stainless steels, superalloys, refractory metals, titanium, and other specialty alloys.

Typical Gas-Quench Vacuum Furnace Operating Cycle

Note: Do not attempt to process an oily and/or wet workload. It cannot be stressed strongly enough that the workload must be clean and dry. An **oily and/or wet workload could result in damage to both the workpieces and the furnace. Ideally, the workpieces should be cleaned just prior to** process-

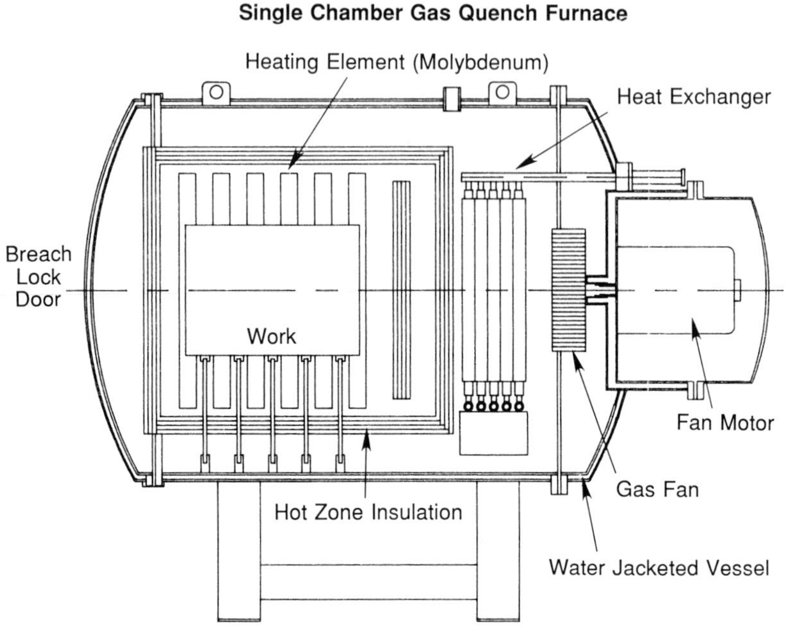

Single Chamber Gas Quench Furnace

Heating Element (Molybdenum)
Heat Exchanger
Breach Lock Door
Work
Fan Motor
Gas Fan
Hot Zone Insulation
Water Jacketed Vessel

Fig 9.3 Section of gas quench vacuum furnace

ing, particularly where humid conditions prevail. Hot-vapor degreasing or an alkaline-water wash followed by hot-air drying are recommended.

1) Set up the workpieces using appropriate fixturing as required. For process temperatures below 925 °C (1700 °F), standard stacking baskets and fixturing made from materials such as RA85H, RA330, Inconel 600, Inconel 601, Inconel 800, and the cast heat-resistant alloys are commonly used. For temperatures greater than 925 °C (1700 °F), Haynes 230 Alloy or Inco MA 956 are an excellent choice. Molybdenum is normally used for temperatures greater than 1205 °C (2200 °F). Graphite or ceramic are occasionally used for fixturing. However, possible reactions with graphite, particularly at elevated temperatures, and failure due to thermal shock with ceramic must be considered. It is important to include the weight of the baskets and fixturing when determining the weight of the workload to be processed and to observe the maximum weight recommendations of the furnace manufacturer.

2) Load workpieces into the furnace, placing a graphite tooling or metal grid on the hearth rails and taking care not to damage the heating elements. If graphite tooling is used, make sure it is insulated from any metal hearth components. Attach at least two thermocouples to the load. Expendable Type K thermocouple wire is a cost-effective way of instrumenting the load provided it is re-tipped or replaced every cycle. Ideally, load thermocouples should be located on the surface of a workpiece and in a location representative of the heaviest section. Too many load thermocouples cannot be used. On occasion, with critical work, up to twelve thermocouples are used, and more commonly two to four. Wipe down the door seal surface and check for O-ring damage.
3) Set the over-temperature control at no more than 28 °C (50 °F) above the maximum temperature to be used.
4) Evacuate the furnace. Often backfilling it with either nitrogen or argon to 15 in. Hg after the vacuum level stabilizes, and evacuating again, will improve the ultimate vacuum and decrease the time required before the cycle can be started. Normally a minimum vacuum level of 1 to 200 microns is required. The vacuum level required depends on furnace construction. From this point on, furnace operation is usually automatic.
5) Ramp the workload to the preheat or process temperature, watching for smooth transition of power level, vacuum, and thermocouple equalization. Ramping the furnace to the processing temperature rather than unrestricted heating is recommended, to minimize temperature overshoot and the possibility of distortion due to overheating. Typically ramp rates of 8 to 28 °C (15 to 50 °F) per min. are used.
6) Soak the workload at the process temperature as required. It is important to note that the workload will lag the set-point by some margin even if a relatively slow ramp rate is used. Time at temperature is usually initiated when all the load thermocouples have reached the set-point temperature tolerance, which can vary from ±6 to 14 °C (±10 to 25 °F).
7) At the end of the soak time the load is quenched, usually automatically by the furnace. The quench sequence normally consists of backfilling the heating chamber rapidly to 20 in. Hg with either nitrogen, argon, or a blend of helium and argon, at which point the gas-quench fan is turned on. Backfilling is continued until the quench pressure is achieved. Some overshoot of the backfill pressure set-point may occur when a high-pressure surge tank is used. Note: Under no circum-

stances must the maximum quench pressure of the furnace be exceeded.

8) There are many different cooling cycles possible, particularly with furnaces which have sophisticated programmable controllers.

   Normal cool. Fan cool at the fastest rate possible for the furnace. Often different pressure set-points are possible.

   Controlled cool. Programmed cooling adding heat or cooling as required by the program. This method is sometimes used when processing large tools so as to avoid excessive thermal gradients within the tool which can lead to distortion or cracking.

   Furnace cool. Remove all heat power and cool in either vacuum or in a partial pressure at the furnace's natural rate of decline for the load.

9) When the cycle is complete, normally at 150 °C (300 °F) or lower, the furnace is either backfilled or pressure is relieved to atmospheric pressure and unloaded. It is important to remember that the furnace is full of a gas which does not support life. Two people should always be present when a furnace is being unloaded, particularly with argon, which is heavier than air.

The workpieces should be cooled to 150 °C (300 °F) or below before tempering. In many instances they must be cooled to below room temperature in a deep freeze. The temper cycle may be carried out in the same furnace that hardened the workpiece, although it is more economical to transfer the work to a lower operating cost atmosphere, air, or vacuum tempering furnace. A vacuum furnace used for hardening is capable of performing a low-temperature temper cycle but it is a less efficient piece of equipment at the lower end of the temperature range, and therefore takes a long time to reach and maintain thermal equilibrium.

It is advisable to check the as-quenched hardness at room temperature before proceeding to the temper, since it will indicate the appropriate tempering temperature to achieve the desired final hardness. The choice of the type of tempering furnace may depend on the temperature at which the work will be processed. Most tool and high-speed steels are usually tempered in the range of 150 to 650 °C (300 to 1200 °F). Many materials may require a double temper, to enhance toughness and other properties.

The quenching gas used is usually nitrogen or argon from a cryogenic source, at purities indicated by the applicable specifications. Nitrogen will provide a faster quench and is less costly, but it cannot be used to quench all materials to be encountered with the vacuum furnace (e.g, titanium). Argon is more costly, but universal in its application, and when purchased in bulk form it is usually economical enough to justify the added expense, which is offset by process safety. Occasionally a mixture of helium and argon is used. The primary advantage of this mixture is that heat transfer is greatly increased as compared to pure argon.

Tool and High-Speed Steels

Many types of tool and high-speed steels can be successfully processed in a gas-quench vacuum furnace. The speed of quenching is not as critical with some types of tool steels as it is with others. Some require cooling rates easily achieved at ⅔ to 1½ atm of pressure with gas fans of modest horsepower. These are the air-hardening tool steels and this group includes the more highly alloyed AISI A, D, S and H series of tool steels.

More severe cooling rates necessitating higher quench pressures of 2 to 6 atm and higher horsepower fans are required to satisfactorily harden larger tools made from these air-hardening tool steels, the leaner AISI O and P tool steels, and the M and T series of high-speed steels. It is important to note that while satisfactory hardness can generally be obtained using a marginal cooling rate, toughness and secondary hardening suffer. Therefore, overall tool quality can suffer. Generally, furnaces with metallic elements do not require as high a pressure and/or as high a horsepower fan as do furnaces with graphite elements, to achieve a comparable cooling rate, because of the lower heat storage of furnaces with metallic elements.

The basic requirement of all tool steels is that they be hardened so as to achieve a long service life with good toughness and/or wear characteristics. The time temperature cycle used is critical to achieving optimum performance, and it varies greatly with the steel being processed and the capability of the furnace used. The most common hardening cycle for A and D series tool steels is preheating at 790 to 815 °C (1450 to 1500 °F) and austenitizing, usually between 925 and 1010 °C (1700 and 1850 °F) for approximately 30 min. at temperature, followed by the most rapid quench possible. The specific temperatures used depend on the ultimate properties desired, whereas the time held at temperature is governed primarily by the section size of the workpieces and the mass of the workload. Typically, a minimum of 30 min. at temperature for each 1 in. thickness of the heaviest cross section of the workpiece is used. At temperatures less than 760 °C (1400 °F), the soak times must be increased as heat transfer is reduced, whereas at temperatures greater than 1095 °C (2000 °F) soak times should be reduced to avoid grain coarsening. Details of these variables may be found in Refs 1 and 2 and in technical literature available from the tool steel producer, which can be a valuable source of this kind of information.

Stainless Steels

The common material categories in this group are the austenitic stainless steels (300 Series) and the martensitic stainless steels (400 Series). The austenitic steels can be hardened only by cold working, not by thermal processing. But they are often processed in vacuum for stress relief and annealing purposes. Unlike carbon steels, which may be stress relieved up to 650 °C (1200 °F), the austenitic series are stress relieved/annealed at temperatures between 1010 and 1120 °C (1850 and 2050 °F), there being very little difference between stress relief and anneal except for the time at temperature.

Austenitic Stainless Steels. Except for the stabilized 321 and 347 grades, the 300 Series require rapid gas-quench to prevent carbide precipitation in the grain boundaries and loss of corrosion resistance if cooled too slowly through the sensitizing range 900 to 425 °C (1650 and 800 °F). The stabilized grades may be furnace-cooled without detrimental effects, but there may be no practical reason to do so other than to conserve quenching gas, or to eliminate distortion in critical parts caused by a severe gas quench.

Martensitic Stainless Steels. On the other hand, the martensitic stainless steels (400 Series) are heat-treat hardenable. The common alloys in this group are the 410, 416, and 440C grades.

The 400 Series stainless steels are hardened in the temperature range of 925 to 1065 °C (1700 to 1950 °F) and must be rapidly gas-quenched to below 205 °C (400 °F) to obtain the as-quenched hardness of 38 to 45 HRC. Again, exact temperature and time are dependent on properties desired in the finished workpiece. Although distortion during quenching is usually not a problem with most gas quenching, since it is considerably less severe than liquid oil quenching, some compromises may have to be made with slow quench rates for workpieces with very thin sections or those with severe thick-to-thin transition areas. Since the 400 Series permit substantially slower cooling rates than alloy steels, this is not a problem.

The martensitic grades (400 Series) require closer attention to heat treating parameters to achieve specific results, as compared with carbon and low-alloy steels. This is true particularly in the high hardness range when tempering, since the slope of the hardness curve is quite steep (i.e., small changes in temperature greatly affect final hardness).

In the high-carbon grades of martensitic stainless steels such as 440C, rapid quenching still leaves high levels of untransformed austenite, which may cause embrittlement or distortion in service. For this reason, sub-zero cooling at –75 to –85 °C (–100 to –120 °F) is often applied for 2 to 16 h followed by double-temper cycles with a 3-hour sub-zero treatment in between. The sub-zero treatment should commence promptly after the completion of the quench from the austenitizing temperature, and the workpiece should be returned to room temperature between the tempering cycles.

There are several annealing cycles for the 400 Series, which range from stress relief to full anneal. These processes are variously called process or sub-critical annealing, isothermal annealing, and full annealing. The temperature to which the work is reheated has a substantial effect on structure and properties. Careful analysis of desired results should be considered before an annealing process is selected. Details of these variables may be found in Refs 1 and 2, and in technical literature available from the tool steel producer. In most cases, slow cooling is recommended, and when a vacuum furnace is used, controlled cooling is readily achieved for the full anneal process.

High-Strength Alloys or Superalloys

The superalloys, sometimes known as high-strength alloys, were developed in large part to satisfy the needs of the aircraft gas turbine industry for higher operating temperatures in the hot sections of jet engines. Generally, those that are heat treatable are precipitation-hardening alloys developed for specific properties in particularly critical gas turbine applications.

Not difficult to heat treat, this group of alloys is very dependable for producing consistent results, provided processing schedules are adhered to strictly. Processing to these precise heat treating schedules became practical with the development of modern vacuum equipment, which allows the user to control all parameters of a thermal cycle to achieve desired properties. The vacuum furnace user can precisely control the rate of temperature rise and the time at the required temperature or any intermediate temperature point, as well as the cooling rate. All these functions can be readily controlled and programmed so that the furnace operator has only to monitor performance, make adjustments for unusual situations, and load and unload the furnace. This is in sharp contrast to pre-cold-wall vacuum equipment, with which the furnace operator normally controlled all parameters by manual operation of the furnace controls.

It is perhaps confusing to the non-metallurgist to understand the hardening process of the precipitation alloys, since it is opposite to that of steel alloys. Steel alloys are hardened by heat treating to a high temperature, and then quenching rapidly to room temperature, at which point they are at maximum hardness. They are then tempered, to slightly reduce hardness but add toughness and other properties. Conversely, precipitation alloys are heated to an elevated temperature to soften them, due to grain structure alignment, and cooled to room temperature, at which point they are soft. They are subsequently hardened by heating to an intermediate temperature where they precipitate constituent elements, typically copper, to their grain boundaries, which results in hardening.

The common language for these processes is solution treating, which is a softening process, or solution annealing, or just annealing. The hardening process is called aging, because it usually requires a number of hours to complete the desired precipitation to the grain boundaries. There is a substantial number of superalloys, which may be found in more detailed technical literature, such as Ref 3, which are processed in this manner. The most common and widely used alloy in this group is Inconel 718, in either wrought or cast form. This material is typically solution annealed in the 940 to 995 °C (1725 to 1825 °F) range, held for 30 min. or longer, then gas-quenched to room temperature. In this condition the material is soft and can be easily worked.

There are two basic hardening or aging cycles, which may achieve the same final hardness in the range of 38 to 48 HRC, but varying other properties, depending on their application. These are the well known "short" and "long" Inconel 718 aging cycles. In many cases economics plays an important role in process selection, because of the higher cost of longer furnace time. The difference is readily apparent as follows:

- The "short" aging cycle is run at 760 °C (1400 °F) for 5 h, slow cooled in the furnace at 56 °C (100 °F) per h to 650 °C (1200 °F), held for 1 h, and then further cooled to room temperature at any suitable rate, for a total furnace time of 10 h, including heating and cooling time.
- The "long" age cycle is run at 725 °C (1325 °F) for 8 h, slow cooled in the furnace at 56 °C (100 °F) per h to 620 °C (1150 °F) and held for 8 h, then further cooled to room temperature by any convenient means, for a total furnace time of 20 h.

Other variations to this aging cycle are less common and result in unique properties.

Titanium

The processing of titanium and its common alloys by the heat treater is usually limited to stress relieving and annealing. The most typical alloy encountered by the vacuum heat treater is the alpha-beta alloy, Ti-6Al-4V, used for its toughness and high strength. Titanium is normally heat treated at some point after complete or partial fabrication, to reduce its residual stresses resulting from prior work. These treatments are usually carried out in vacuum at temperatures in the range of 480 to 760 °C (900 to 1400 °F), depending on the alloy and the final properties desired. Soaking times at these temperatures are 1 to 2 h, and the work may be gas-fan cooled. Although the rate of cooling for stress relieving temperatures is not critical, uniformity is particularly important down to 315 °C (600 °F). The vacuum furnace is ideal for this work, since it is advisable not to employ water quenching to accelerate cooling as it tends to induce additional stresses because of unequal cooling between the surface and center of the work.

Solution Treating. Solution treating is usually carried out by the mill supplier, in the case of alpha-beta alloy Ti-6Al-4V, at approximately 900 to 955 °C (1650 to 1750 °F), with a water quench to achieve the cooling rates necessary for the required properties. Because quench-delay time is extremely critical in this alloy, 7 s or less of water or brine quenching is the only option and therefore the possibility of vacuum furnace processing is eliminated. Other alloys are less sensitive to quench-delay time, but workpiece size and geometry may also rule out gas-fan cooling.

Aging or Stabilizing Treatments. Aging treatments usually follow solution treating and may require anywhere from 4 to 24 h furnace time at low temperatures of 425 to 760 °C (800 to 1400 °F). They may be gas-fan cooled in a vacuum furnace, but since solution treating is done by the mill, they usually follow with the aging process prior to fabrication.

Although solution treating and aging are normally carried out in non-vacuum furnace equipment, it is before fabrication and a considerable amount of metal may be removed in cold working. Therefore, surface condition is not a primary concern. However, the stress relieving annealing cycles are processed in vacuum, because the workpiece is either at or close to final dimension and surface condition is critical.

Heat Treating. The heat treating of titanium alloys is a complicated process and in the case of the alpha-beta alloy Ti-6Al-4V, the exact thermal process schedule in vacuum will become a series of compromises and modifications to achieve the desired properties and service requirements.

In heat treating titanium and its alloys, the surface condition (structure) is modified because titanium reacts with water, oxygen, hydrogen, and carbon dioxide normally found in varying degrees in all heating environments. This typically oxygen-enriched layer produces the unique metallurgical disease called "alpha case," a brittle layer that must be removed prior to service of the workpiece, since it may seriously reduce mechanical properties. It is also called "white layer," and it may vary from as little as millionths to thousandths of an inch in thickness. Major titanium users put limits on the amount of alpha case they will allow, if any. This layer is usually removed by machining or chemical etching after thermal processing.

The precleaning of titanium prior to heat treatment is of primary concern so as to minimize alpha case. Caution must be exercised in the use of cleaning materials as well as the removal of any loose surface contamination, since undesirable reactions will take place on the surface at the elevated temperatures at which the titanium is processed. Contaminants such as oil, grease, fingerprints, and water, must be removed and the workpiece dried prior to heat treatment. Titanium workpieces are never cleaned in a typical hydrocarbon degreasing medium, since embrittlement is likely to occur. Workpieces may be washed in distilled water (ordinary tap water should be avoided), or cleaned with acetone or another similar agent. Fixtures and tooling must similarly be free of any foreign surface material. Once cleaned, titanium workpieces should be handled only with clean white gloves.

Alpha case is generated to some degree in almost all environments at elevated temperatures, but selection of the furnace environment will determine the degree to which it develops. Therefore it is generally agreed that the use of a low-leak-rate vacuum system (10 microns per h or less) is preferred and that the resulting brittle layer will be almost negligible or undetectable by metallographic means, provided test specimens are run with the work load for verification.

Key considerations in heat treating titanium and its alloys, practices that are to be followed, and those that should be avoided, are summarized below, from Ref 4:

- Provide sufficient stock for post-treatment metal removal requirements (contaminated metal removal) where possible
- Clean components, fixtures and furnace prior to heat treatment. Caution: Do not use ordinary tap water for cleaning titanium components
- Use temperature controls with separate over-temperature protection to ensure temperature does not exceed beta transus
- Stack and support components to allow free access of heating and quenching media
- Observe quench-delay requirements to ensure hardening response during aging
- Review property requirements and select optimum heat treating procedure
- Review strength requirements and select proper aging cycle

- Remove alpha case after all heat treating is complete
- Check for the presence of hydrogen after all processing is complete
- Do not nest components
- Do not allow temperature to exceed beta transus (unless it is specified as a beta anneal process)
- Do not rely on inert atmosphere or vacuum for prevention of oxygen contamination
- Do not rely on hardness tests for measurement of the effects of heat treatment
- Do not pickle assemblies with faying surfaces

Specialty Alloys

A number of specialty alloys are encountered by the vacuum heat treater, each designed to accomplish a specific function or application. Many of these alloys were previously processed in a reducing or protective atmosphere before vacuum was available, and many specifications are written based on the use of dry hydrogen, dissociated ammonia, inert gases, or neutral atmosphere. In most cases, vacuum is a very practical, cost effective alternative and is now in common use for these materials. Typical of the more common specialty alloys are Mumetal and beryllium copper.

Mumetal. Mumetal is a magnetic alloy, because of its magnetic shielding capability when handled and heat treated properly. It is one of the common nickel-iron based alloys used in a variety of instruments as a shield for stray magnetic fields. Its basic heat treating cycle is 1120 to 1150 °C (2050 to 2100 °F) for 4 h, followed by a slow, controlled cool at 56 °C (100 °F) per h or less, to below 40 °C (100 °F). Cleanliness prior to heat treating is very important. Workpieces should be cleaned and handled with clean white gloves prior to heat treatment. Otherwise, undesirable fingerprints will mar them after processing, because the hydrocarbons are driven into the surface, forming a perfect pattern. Workpieces must not touch each other during heat treating, as they will bond together by diffusion bonding. After heat treatment, workpieces must be handled with extreme care. The material is very soft and any bumping or impact will tend to distort the aligned grain structure, resulting in a loss of shielding properties.

Beryllium Copper. Beryllium copper is used mostly for instrument parts and other small devices. It is easily machined and a single vacuum heat treatment age hardens the material at a low temperature, but maintains surface cleanliness rather than excessive oxidation and scale when treated in air. Because beryllium copper is used to make small, delicate instrument parts, the use of vacuum processing eliminates the need for additional cleaning, because it produces a bright, clean surface. Alternative means of producing such a surface require the use of hydrogen which, if not handled carefully, can be dangerous.

The most common heat-treat process for beryllium copper is aging, which hardens the material by the metallurgical function of precipitating other elements within the grain structure of the copper. Tables are available with proper times and temperatures for selected properties, but in general 315 °C (600 °F) for 2 h is the normal cycle.

Brazing

Vacuum brazing has become the pre-eminent process form for joining most materials by the brazing technique, as defined by the American Welding Society:

"Brazing is a group of welding processes which produce a coalescence of materials by heating them to a suitable temperature and by using a filler metal having a liquidus above 450 °C (840 °F) and below the solidus of the base material. The filler metal is distributed between the closely fitted surfaces of the joint by capillary attraction."

The modern vacuum furnace has made it possible to join numerous materials heretofore difficult or impossible to braze in other environments. It has led to the development of special techniques, improved processes, and faster cycle times in furnace brazing. Much of this progress must be credited to the defense efforts of World War II, and the development of the aircraft gas turbine and subsequent aerospace missiles. Today, most materials are routinely vacuum brazed using the batch vacuum process with modifications in system pressure levels to reduce or eliminate the outgassing of some of the higher vapor pressure elements of filler metals at brazing temperature.

Accordingly, there are three basic filler metal groups of general concern, each used to braze a variety of base metal brazements:

1) Silver-base filler metals
2) Copper-base filler metals
3) Nickel-base filler metals

For both the silver- and copper-base metals it is advisable to adjust the system vacuum level to what is commonly called a "partial-pressure mode," or a pressure level less than atmospheric but higher than what is considered a high vacuum of 10^{-3} torr or less (1 micron). It is a pressure somewhere in between, usually 300 to 1000 microns vacuum, and typically 600 microns. Brazing at these partial-pressure levels not only inhibits outgassing and loss of filler metal elements at the higher temperatures due to the vapor pressure of those elements, but it allows all the benefits of vacuum treating, such as uniform and controlled heating and cooling rates, cleanliness, fast thermal response, and short cycle times.

Silver-Base Filler Metals

The silver-base family of filler metals is represented by the more common furnace braze alloys designated by the American Welding Society (AWS) as Bag-8, Bag-13, and Bag-18. These alloys of silver and copper have small additions of nickel, tin, and zinc to adjust their melting point and brazing range, to allow their use in connection with heat treating the base metal coincident with the braze process, and to enhance other mechanical properties. In most cases these filler metals can be used effectively without flux. However, in some situations the addition of small amounts of flux is beneficial, consistent with the possibility that some vacuum system cold-zone contamination may result. It becomes an economic trade-off, but as a general rule flux should not be used in the vacuum system, because it may condense on cool surfaces inside the furnace and contaminate subsequent loads.

Figure 9.4 shows a typical silver-braze thermal cycle. Work is loaded onto the furnace hearth and the system temperature is raised at a rate of 11 to 17 °C (20 to 30 °F) per min. to 10 to 40 °C (50 to 100 °F) below the liquidus of the filler metal being used, commonly called the "hold off" temperature. It is held at this temperature long enough to allow a complete soak of the braze joint area and filler metal, usually from 10 to 30 min., depending on factors such as load size, geometry, and complexity of the workpiece.

Full power is then applied to the heating elements to raise the temperature to the filler metal liquidus and/or braze range as rapidly as possible, to avoid liquation of the filler metal elements. This rapid rise to braze temperature is one of the most important steps of the braze process, when braze filler metals with more than two constituents are used. Liquation, or separation of a filler metal results in distortion of the true braze temperature and produces an incomplete and unsatisfactory braze joint. Lack of braze joint integrity is often the result of this type of improper processing. Once braze temperature has been attained, unless there are concurrent heat-treat requirements, temperature should be held for 5 to 15 min. for the typical braze joint. Obviously, there are many exceptions to this rule, such as for very thin metal sections, where 2 min. may be sufficient to assure complete and

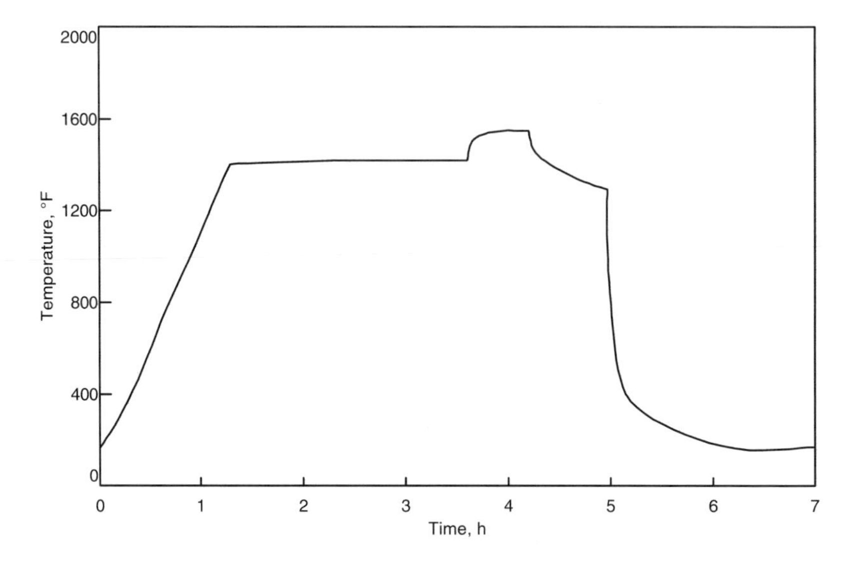

Fig 9.4 Silver braze thermal cycle

uniform heating, or at the other end of the time window, for long joints in heavier sections. As will be noted later, time at temperature may be dependent not only on the type of workpiece being brazed but also on the filler metal alloy.

Time at temperature is terminated by shutting down power to the heating elements. The workpiece temperature will immediately drop off sharply, particularly at the higher braze temperatures above 980 °C (1800 °F), provided the workload is not massive. To assure complete solidification of the filler metal in the braze joint, and for other metallurgical considerations, it is advisable to furnace-cool the work to several hundred degrees below the braze temperature, or at least below the braze filler metal solidus before initiating the quench sequence, and to cool it to below 150 °C (300 °F) before opening the furnace. This avoids discoloration of the workpiece and possible oxidation of some furnace components.

Copper-Base Filler Metals

Copper-base filler metals are essentially pure copper, with additives to create special alloys for unique purposes. The partial-pressure vacuum mode must be used, to avoid furnace contamination by vaporized copper deposition on furnace cold-zone areas. This is an important factor, because the basic copper braze temperature is 1120 °C (2050 °F), far above the average silver-base filler metal braze temperature of 760 to 870 °C (1400 to 1600 °F). Deposition of copper or other contaminants on the reflective heat shields of the furnace seriously reduces their thermal emissivity, increases power consumption, reduces overall efficiency, and makes the furnace more difficult to control.

Brazing with copper has one unique feature that makes it stand alone in the overall family of braze alloys: its ability to flow through or "pull" great lengths of braze joints in carbon steels, provided a zero-clearance or interference fit is presented. Molten copper has been demonstrated to travel a braze joint of 48 in. under such conditions, when available in adequate quantities. This occurs because pure copper has only one melting point, approximately 1080 °C (1980 °F), has a low viscosity in the liquid state, and responds readily to capillary action under the proper braze-joint conditions.

Nickel-Base Filler Metals

Nickel-base filler metals are vacuum-brazed in full vacuum of 1 micron (10^{-3} torr) or less. They are used primarily for brazing stainless steels and high-strength superalloys, for both aerospace and industrial applications. They have liquidus and brazing temperature ranges above 980 °C (1800 °F), higher strengths often equalling the base metals, and corrosion resistance; but they also have some unacceptably high as-brazed hardness values, and therefore suffer from reduced ductility.

Nickel-base braze cycles are similar to those of the silver- and copper-base filler metals, but often a hardening, austenitizing, or solution cycle is combined with the braze cycle, followed by single- or double-temper cycles, or precipitation-aging cycles in the case of the higher-strength superalloys. The additional time required by the aging cycles has no effect on the braze joint except for perhaps some minor additional diffusion of the filler metal into the faying surfaces of the joint, because precipitation takes place at a far lower temperature than the filler metal solidus.

In the case of nickel-base filler metals, flux is rarely used, especially since the thermal cycles are carried out under full vacuum. Because of this, post-braze cleaning is never a problem, and often finished brazed assemblies require no further processing. Most of the filler metals in the group have wide melting ranges (the temperature range between the solidus and liquidus of the alloy, due to the multi-element features of these materials) which allow the modification of specific cycles used with a particular filler metal alloy to achieve or overcome unusual problems.

For example, it is possible (and not uncommon) to braze at the lower end of the brazing ranges of a particular filler metal alloy so as to fill an unusually large braze gap, where no other options are available other than to add foreign filler materials (which produces a marginal braze joint), or to braze at the high end of the brazing range (to achieve a heat treating requirement), or to "pull" a longer joint. Ideal joint clearances for this category of filler metals vary between 0.0254 to 0.127 mm (0.001 to 0.005 in.). However, the brazing source is often called upon to braze gaps in excess of 0.254 mm (0.010 in.) by the use of fillers such as pure nickel powder.

Although erosion of base metal by the filler metal, due to extended time at temperature in this alloy group is usually not a problem, there is one common exception. The braze filler metal known as "gold/nickel," BAu-4, (82% gold, 18% nickel), is a eutectic alloy with only one melting temperature (solidus and liquidus are equal). Cost notwithstanding, it is one of the highest-quality filler metals available, with superior strength and corrosion resistance, and excellent ductility. It has a unique feature which not only sets it apart but requires caution in its application: a tendency for erosion of the base metal if left in the joint in the liquid state for too long a period. Braze time at temperature should be limited to no more than 5 to 8 min., and in the case of very thin workpieces should be reduced to 2 to 3 min. Erosion of base metal in the braze joint not only weakens the material but also negates most of the other advantages of this premier filler metal.

Titanium and Aluminum

Titanium vacuum brazing has become a unique field in itself, not unlike the special discipline of aluminum brazing in vacuum. Each requires special considerations in furnace equipment design and performance. Vacuum aluminum brazing has developed into a high-quality, high-volume, production process, primarily in heat exchanger applications. Furnace equipment is usually designed to braze a specific geometry, often on a continuous basis, as in the case of radiator cores. Such equipment is not readily adapted to a variety of other small-quantity workpieces, and therefore this process has not generally developed as an all-purpose brazing system. Titanium vacuum brazing has a particular niche in the aerospace industry due to the increased application of this base metal to aircraft and gas turbine engines.

Furnaces for vacuum brazing of titanium are of the same basic type used for the other filler metals, except that they require better and more sophisticated internal components, higher vacuum, and minimal leak rates (5 microns per h or less). Vacuum levels should be in the 10^{-5} to 10^{-6} torr range and often the addition of getters, to ensure adequate degassing, is required to achieve satisfactory results.

Thermal cycles are similar to nickel-braze cycles except that they are run at lower temperatures as with silver alloys. Both silver-base and aluminum filler metals are in common use for brazing titanium. As in the case of vacuum aluminum brazing, process techniques tend to be proprietary in titanium brazing, and much is still to be exposed to the general vacuum industry.

Sintering P/M Material Systems

Vacuum sintering of P/M materials has been a commercially accepted manufacturing process for over 30 years. This process was first developed for sintering refractory and reactive metals such as niobium, tantalum, titanium, zirconium and their alloys, but it has been extended to include stainless steels, tool steels, and low-alloy steels. Vacuum sintered P/M products include bulk forms, near-net-shape parts, conventional P/M parts, and metal injection molded (MIM) components.

Vacuum sintering offers the P/M parts manufacturer a means of sintering materials at higher temperatures than those possible with conventional metal mesh belt furnaces, which are typically limited to an 1150 °C (2100 °F) maximum operating temperature, because of the reduced strength of normal belt materials above this temperature. Vacuum systems extend the useful sintering temperature range well beyond this limit. In addition, problems encountered in sintering with reactive atmospheres can be eliminated by using a vacuum environment, and it also helps to remove absorbed gases and volatile oxides found on the powder particle surfaces, thereby accelerating the sintering process.

Stainless Steel

The physical and mechanical properties of P/M stainless steel alloys are affected by the sintering environment. To achieve the best corrosion resistance, the residual carbon, nitrogen, and oxygen levels should be as low as possible. For vacuum sintering a backfill gas is required to control vaporization of chromium from the surface of the product. Typically nitrogen or argon is used as the backfill gas, at a pressure of 400 to 600 microns. If nitrogen is used, the cooling rate from sintering temperature must be high, 70 °C (125 °F) per min. or greater, to prevent formation and precipitation of chromium nitrides in the alloy. Formation of this nitride phase will deplete the alloy of chromium and lower its corrosion resistance. To eliminate any problems with nitride formation the use of argon is recommended. Higher sintering temperatures will also help produce the maximum corrosion resistance.

With regard to mechanical properties, nitrogen dissolved in the matrix of austenitic stainless steels such as 304 or 316, will increase yield strength but reduce ductility and impact resistance. Selection and control of the backfill gas during vacuum sintering can have a profound effect on the properties of P/M stainless steels.

Soft Magnetic Alloys

Several soft magnetic alloys, such as Fe-0.45P, Fe-3DSSi, and Fe-50Ni, are vacuum sintered. Because optimum magnetic properties require the lowest level of residual impurities (carbon, oxygen, nitrogen, sulfur), argon is recommended as the backfill gas for vacuum sintering these alloys. In addition, the highest possible sintering temperature should be used to accelerate grain growth. Large grain size in these alloys also improves magnetic response.

Tool Steels

Full-density P/M tool steels are routinely produced by vacuum sintering. Transient liquid phase formation during sintering accelerates densification and eliminates almost all residual porosity. To achieve full density, temperature control must be extremely precise: a variation of ± 3 °C (±5 °F) is normally required throughout the furnace load. Under-sintering will produce parts at less than full density; over-sintering will cause excessive liquid phase formation with massive carbide formation and slumping of the workpieces.

Titanium

Because of the reactivity of titanium alloys, vacuum sintering of them requires a high vacuum normally achieved with a mechanical pump plus diffusion pump system. Vacuum sintering removes any gases trapped in the structure, provides for liquid phase formation (for such alloys as Ti-6Al-4V), accelerates sintering, and leads to near-full density products.

Cemented Carbides

In sintering tungsten and other cemented carbides, they are cemented by a liquid cobalt phase. Vacuum furnace sintering has replaced atmosphere furnace sintering for these materials because it makes higher temperatures easier to attain and the sintering environment easier to control. With the development of special low temperature controls and wax collection systems, it is possible to remove the paraffin wax binder and sinter in the same furnace. Effective condensing systems are necessary to trap the paraffin vapor before it reaches the pump, as the presence of paraffin may impair the lubricating qualities of the pump oil. As about 50% of the sintering cycle is spent reaching the desired vacuum and temperature, and the balance of the cycle is used for cooling, batch furnaces are often connected in pairs to a common vacuum pumping system, power supply, and controls. Continuous vacuum sintering furnaces for cemented carbides have a rotating table on which the material is dewaxed, sintered, and cooled in a vacuum. In a horizontal continuous vacuum furnace the product is dewaxed in partial-pressure hydrogen and sintered in a vacuum, with vacuum interlocks between the dewaxing and the high-heat chamber, and also between the high-heat chamber and the cooling chamber.

Lubricant/Binder Removal

Most P/M parts contain a lubricant or binder in the green compact, which must be removed prior to vacuum sintering to keep the vacuum pump oil clean. When paraffin wax is used as a binder, special cold traps built into the vacuum pumping system can be used to remove the binder phase vapors in a similar manner as for cemented carbides. For the majority of vacuum sintered parts the lubricant is burned off in a forced-air or controlled-atmosphere furnace just prior to vacuum sintering. Synthetic wax lubricants are preferred over metallic stearates for vacuum sintering, to prevent residual metallic vapors from contaminating the vacuum furnace. Air burn-off is typically used to remove lubricants from stainless steel products. When carbon control is necessary, as for low-alloy steels or tool steels, a neutral or reducing burn-off atmosphere is required, to prevent loss of admixed graphite. If one or more of the powder constituents forms a tenacious oxide, such as ferrosilicon or titanium, lubricant removal must be performed in a reducing atmosphere.

Fixturing

Workpieces to be sintered must be arranged on fixtures, to fully load the vacuum furnace chamber. Because of its light weight and high-temperature strength, graphite trays, boats, and plates are used as fixtures for vacuum sintering. Since the workpiece material in many cases can react with the graphite, a ceramic coating or layer is used to separate it from the graphite fixture. An alumina wash can be used to completely coat the fixture, or in some cases a thin alumina plate or alumina paper or felt is used to insulate the workpiece from the graphite. The trays or plates are stacked, using graphite or ceramic separators, and loaded into the furnace on a graphite hearthplate. In some cases, special shielding may be used to prevent direct radiation from the heating elements from impinging on the product.

Furnace Operation

After the workload (or product) has been loaded into the vacuum furnace, the chamber is evacuated to the desired vacuum level. The heat is turned on and the chamber further evacuated during outgassing of the work. Heating continues at an established rate to an intermediate hold temperature. The backfill gas is introduced and after the hold period the load is heated to final sintering temperature. More than one intermediate hold and different heating rates are used, depending on the material being sintered. After holding at the sintering temperature for the required time (30 min. to several hours) the heat is turned off. The chamber is backfilled to nearly 1 atm pressure, and the cooling fan is turned on to accelerate the cooling cycle. In multi-chamber furnaces the product can be removed from the heating chamber into a separate cooling chamber when more rapid cooling is desired. Finned, water-cooled heat exchangers in the cooling chamber assist in heat transfer and more rapid cooling.

Case Histories for Oil-Hardening Metals

Low-Alloy Steel

Workpiece: Sleeves, liners, and retainers for Army helicopter transmission. Various configurations, typically thin-wall flanged cylinders forged from AISI 4340ESR steel, normalized, rough-machined, hardened, tempered, and finish-ground.

Section thickness: 6 mm (approximately 0.25 in.)

Required mechanical properties: 285 to 305 ksi ultimate strength, 54 to 57 HRC.

Loading arrangements: Mesh-lined stacking baskets with flat bottoms to ensure adequate workpiece support. Stainless steel expanded-metal screen on top of each layer of workpieces in baskets to permit additional layers as required. Tensile bars oriented vertically, held loosely in place in each basket with stainless steel wire, care taken not to over-tighten, which could distort tensile bars.

Furnace hearth cart: Inconel 600 or Haynes 230 support bars and stainless steel heat shields.

Atmosphere: Furnace evacuated to less than 500 microns, backfilled to 15 in. Hg with nitrogen and evacuated again to less than 100 microns.

Thermal cycle: Ramped unrestricted to 260 °C (500 °F), then ramped to 870 °C (1600 °F) at 14 °C (25 °F) per min. Soaked 870 °C (1600 °F) for 3 h with a guaranteed soak band of 6 °C (10 °F). Quenched in well-agitated medium-fast oil at 40 °C (100 °F) under 10 in. Hg nitrogen. Tempered 3 h at 165 °C (325 °F) in air.

Tool Steel

Workpiece: Punch, machined from S5 tool steel, with a heavy base relative to a long (8 in.) thin-wall punch section. Where the two sections meet, cracking has been a problem.

Required mechanical properties: 55 to 60 HRC.

Loading arrangements: Loaded oriented vertically, base down, in stacking baskets with mesh liner.

Furnace hearth cart: Inconel 600 or Haynes 230 support bars and stainless steel heat shields.

Atmosphere: Furnace evacuated to less than 500 microns, backfilled to 15 in. Hg with nitrogen, and evacuate again to less than 100 microns.

Thermal cycle: Ramped unrestricted to 260 °C (500 °F), then ramped to 760 °C (1400 °F) at 11 °C (20 °F) per min. Soaked 760 °C (1400 °F) for 75 min. with a guaranteed soak band of 6 °C (10 °F). Ramped to 870 °C (1600 °F) at 14 °C (25 °F) per min. Soaked 870 °C (1600 °F) for 15 min. with a guaranteed soak band of 6 °C (10 °F). Quenched in well-agitated medium-fast oil at 40 °C (100 °F) under 10 in. Hg nitrogen for 10 s, removed from oil for 20 s, and left in oil until furnace ready to be unloaded. Tempered 2+ h at 175 °C (350 °F) in air.

Oil-Hardening Steels

In all cases, workload and hearth cart hot-vapor degreased prior to loading into furnace and again after completion of oil quench.

High-Speed Steel

Workpiece: Thread-rolling die for forming threads on stainless steel tubing, machined from P/M grade M4 steel, approximately 4 in. diameter by 4 in. high with a 2 in. diameter hole in the center.

Section thickness: 50 mm (2 in.)

Required mechanical properties: 66 to 68 HRC.

Loading arrangements: Loaded in mesh-lined stacking basket so holes are vertically oriented. The basket must have a flat bottom to ensure the workpieces are adequately supported.

Furnace hearth cart: Molybdenum and nickel heat shields, one wheel on each side of the cart blocked to prevent it from drifting on the quench elevator during interrupted quench.

Atmosphere: Furnace evacuated to less than 500 microns, backfilled to 15 in. Hg with nitrogen, and evacuated again to less than 100 microns.

Thermal cycle: Ramped unrestricted to 260 °C (500 °F), then ramped to 1010 °C (1850 °F) at 14 °C (25 °F) per min. Soaked 1010 °C (1850 °F) for 30 min. with a guaranteed soak band of 6 °C (10 °F). Ramped to 1190 °C (2175 °F) at 14 °C (25 °F) per min. Soaked 1190 °C (2175 °F) for 10 min. with a guaranteed soak band of 6 °C (10 °F). Transfer to quench chamber over oil, but quench delayed 30 s. Quenched in well-agitated medium-fast oil at 38 °C (100 °F) under 10 in. Hg nitrogen for 15 s, removed from oil for 15 s, quenched for 15 s, removed from the oil for 15 s, quenched and left in oil until furnace ready to be unloaded. Tempered by deep-freezing at –85 °C (–120 F) for 4 h, warming to room temperature and tempering for 4 h at 525 °C (980 °F), and repeating this deep-freeze/temper cycle four times.

Corrosion-Resistant Steel

Workpiece: Ball bearings of 440C steel, ranging in diameter from 1/8 in. to 3/4 in.

Required mechanical properties: 60 HRC after deep freezing and tempering.

Loading arrangements: Loaded in stacking baskets lined with mesh with appropriate-size openings. Baskets must have properly fitted support bars to ensure that those in middle and upper layers are adequately supported. Stainless steel expanded metal is placed on top of the balls in the top basket. Important that the maximum surface area for quenching not be exceeded. Maximum load of 1/8 in. diameter balls is 170 lbs, but for 3/8 in. balls, 450 lbs. Specific recommendations from furnace manufacturer.

Furnace hearth cart: Haynes 230 Alloy support bars and stainless steel heat shields.

Atmosphere: Furnace evacuated to less than 500 microns, backfilled to 15 in. Hg with nitrogen, and evacuated again to less than 100 microns.

Thermal cycle: Ramped unrestricted to 260 °C (500 °F) and ramped to 1065 °C (1950 °F) at 14 °C (25 °F) per min. Soaked at 1065 °C (1950 °F) for 90 min. with a guaranteed soak band of 6 °C (10 °F). Quenched in well-agitated medium-fast oil at 38 °C (100 °F) under 10 in. Hg nitrogen. Tempered by deep freezing at -85 °C (-120 °F) for 2 h, warming to room temperature, and tempering for 2 h at 150 °C (300 °F).

Case Histories for Gas-Quenching Metals

In all cases, workpieces stacked in groups or individually fixtured to maintain flatness and roundness and minimize distortion, depending on process temperature, time at temperature, heating/cooling rate, part geometry, cross-sectional thickness, and workpiece fragility. These variables required many compromises based on experience and deference to customer requirements and expectations. Temperatures determined when the last thermocouple reaches required temperature, and time at temperature started then. Temperatures stated are ±15 °C (±25 °F).

Vacuum Annealing Hastaloy X (AMS-5536)

Workpiece: Sheet metal cones, approximately 12 in. diameter by 7 in. high, made from 0.040 in. stock (see Fig 9.5).

Loading arrangements: Nested in stacks of 6 pieces, separated by banding iron straps or pieces of ceramic fiber blanket.

Atmosphere: Vacuum level maintained at 10^{-3} torr or higher.

Thermal cycle: Heated at a rate of 17 to 22 °C (30 to 40 °F) per min. to 1165 to 1175 °C (2125 to 2150 °F), held for 10 to 15 min. and rapid gas-fan-cooled in nitrogen or argon to below 150 °C (300 °F) before removing from furnace.

Fig 9.5 Stock nested in stacks of 6 pieces in in-process vacuum annealing

Fig 9.6 Stocked individually on graphite blocks to harden and temper AISI 410 stainless steel

Hardening and Tempering AISI 410 Stainless Steel

Workpiece: AISI 410 stainless steel sheet metal bands, precisely formed to approximately 20 in. diameter by 4 in. high from 0.032 in. stock.

Required mechanical properties: 38 to 42 HRC.

Loading arrangements: Individually stacked, each on 6 graphite blocks, to maintain flatness and minimize their tendency to become oval in shape (see Fig 9.6).

Atmosphere: Vacuum level maintained at 10^{-3} torr or higher.

Thermal cycle: Heated at 17 to 22 °C (30 to 40 °F) per min. to 980 to 1010 °C (1800 to 1850 °F), held for 30 min. and gas-fan-cooled to below 150 °C (300 °F). Tempered by transferring to vacuum tempering furnace at 510 °C (950 °F) for 2 h, and gas-fan-cooling to below 150 °C (300 °F).

Solution Annealing and Aging Inconel 718 Stainless Steel

Workpiece: Forged Inconel 718 stainless steel engine shroud rings, rough machined, approximately 40 in. diameter by 6 in. high, with ½ in. cross section.

Loading arrangements: Workpieces stacked directly on each other, 9 high, alternating to match minor and major flanges (see Fig 9.7).

Required mechanical properties: 38 HRC minimum.

Atmosphere: Vacuum level maintained at 10^{-3} torr or higher.

Thermal cycle to solution anneal: Heated at maximum rate of 28 to 34 °C (50 to 60 °F) per min. to 955 to 980 °C (1750 to 1800 °F), held 60 min. and gas-fan-cooled to below 595 °C (1100 °F) within 23 min., and further cooled to below 150 °C (300 °F).

Thermal cycle to precipitation-harden: Reheated at the same rate to 720 °C (1325 °F), held for 8 h, controlled-cooled at the rate of 56 °C (100 °F) per h down to 595 °C (1100 °F), held for 8 h, and gas-fan-cooled to below 150 °C (300 °F).

Stress-Relief Annealing Ti-6Al-4V

Workpiece: Fusion welded 300 to 400 lb Ti-6Al-4V jet engine inlet cases.

Loading arrangements: In 300 series stainless steel fixtures to maintain dimensional stability (see Fig 9.8).

Atmosphere: Vacuum level maintained at 10^{-3} torr or higher.

Thermal cycle: Heated at 17 to 28 °C (30 to 50 °F) per min. to 620 °C (1150 °F) for 2 h and gas-fan-cooled to below 150 °C (300 °F).

Fig 9.7 Solution annealing and aging of Inconel 718 showing forged rings stacked directly on each other

Fig 9.8 Jet engine inlet cases for stress relief annealing

Age Hardening Beryllium Copper

Workpiece: Beryllium copper small instrument wheels (see Fig 9.9).

Loading arrangements: Workpieces vacuum aged in bulk form on stainless steel trays.

Atmosphere: Vacuum level maintained at 10^{-3} torr or higher.

Thermal cycle: Heated to 315 °C (600 °F), held for 2 h and gas-fan-cooled to below 150 °C (300 °F).

Silver Brazing OFHC Copper

Workpiece: Heat exchanger modules of oxygen-free high-copper copper, with filler material per Bag-8, a silver copper eutectic alloy.

Atmosphere: 600 to 1,000 microns partial pressure.

Thermal cycle: At 845 °C (1550 °F) for 5 to 10 min. and-gas-fan-cooled (see Fig 9.10).

Copper Brazing Carbon Steel

Workpiece: Carbon steel heat exchangers ("Bee-Bee" type commercial "boilers")

Filler material: Pure copper.

Furnace: Hot-wall vertical type.

Atmosphere: 600 to 1,000 microns partial pressure.

Fig 9.9 Instrument wheels in stainless steel trays for age hardening

Fig 9.10 Copper heat exchanger modules

Thermal cycle: Brazed at 1135 °C (2075 °F) for 5 to 10 min. and gas-fan-cooled (see Fig 9.11).

Nickel Brazing Stainless Steel

Workpiece: Stainless steel heat exchanger cooling coils

Filler material: Nickel-base per AMS-4775, applied in paste form.

Stop-off material: Aluminum oxide type, painted on workpiece flange to prevent build-up of braze metal run-off from coils.

Atmosphere: Vacuum level maintained at 10^{-3} torr or higher.

Thermal cycle: Brazed at 1065 °C (1950 °F) for 10 to 12 min. under full vacuum and gas-fan-cooled (see Fig 9.12).

Case Histories for Sintering

Low-Alloy Steel

Material: Fe+2%Ni+0.5%Mo+1%Cu+0.8% graphite + 0.75% Acrawax C

Density: 7.0 g/cc

Burn-off cycle: 540 °C (1000 °F) in dissociated ammonia for 1 h

Sinter cycle: Evacuate vacuum chamber to less than 50 microns; heat to 815 °C (1500 °F) at maximum power, at the rate of 28 °C (50 °F) per min.; hold at 815 °C (1500 °F) for 30 min.; backfill with nitrogen to a partial pressure of 400

Fig 9.11 Carbon steel heat exchangers

Fig 9.12 Nickel brazed heat exchangers

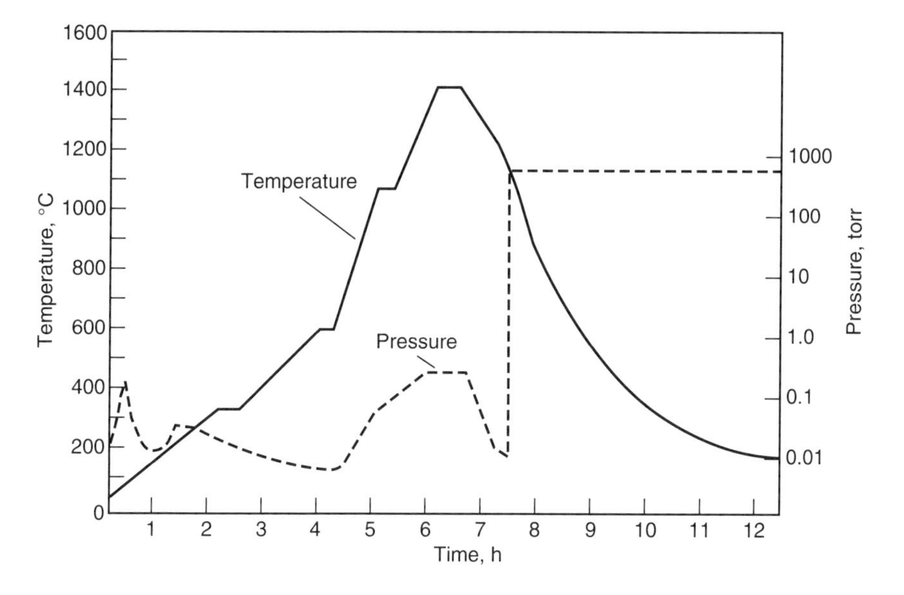

Fig 9.13 Sinter cycle for cemented carbide

microns; heat to 1260 °C (2300 °F), at a rate of 14 °C (25 °F) per min.; hold at 1260 °C (2300 °F) for 1 h; turn off power and transfer to cooling chamber; backfill with nitrogen to a partial pressure of 25 in. Hg; rapid cool.

Final properties: density, 7.15 g/cc; tensile strength, >120,000 psi.

Tool Steel

Material: M-2 high-speed steel + 0.75% Acrawax C

Density: 6.5 g/cc

Burn-off cycle: 650 °C (1200 °F) in N_2/H_2 blend atmosphere for 1 h.

Sinter cycle: Evacuate vacuum chamber to less than 20 microns; heat to 760 °C (1400 °F), at a rate of 28 °C (50 °F) per min.; hold at 760 °C (1400 °F) for 1 h; heat to 1095 °C (2000 °F), at a rate of 14 °C (25 °F) per min.; hold at 1095 °C (2000 °F) for 1 h; backfill with 500 microns of nitrogen; heat to 1215 °C (2225 °F), at a rate of 6 °C (10 °F) per min.; hold at 1215 °C (2225 °F) for 1 h; heat to 1240 °C (2260 °F), at a rate of 0.6 °C (1 °F) per min.; hold at 1240 °C (2260 °F) for 90 min.; turn off power and cool to room temperature.

Final properties: density, >8.0 g/cc; mechanical properties after heat treatment are similar to wrought M2 high-speed steel.

Iron-3% Silicon Steel

Material: Fe-10% Ferrosilicon (31%Si) + 0.75% Acrawax C

Density: 7.1 g/cc

Burn-off cycle: 675 °C (1250 °F) in dissociated ammonia for 1 h.

Sinter cycle: Evacuate vacuum chamber to less than 30 microns, heat to 650 °C (1200 °F), at max power; hold for 30 min.; heat to 980 °C (1800 °F), at max power, hold for 30 min.; backfill to 200 microns with argon; heat to 1175 °C (2150 °F), at max power; hold for 30 min.; heat to 1230 'C (2250 °F), at max power; hold for 2 h; turn off power; furnace-cool to 650 °C (1200 °F); rapid-cool to room temperature.

Final properties: density, 7.3 g/cc; magnetic characteristics, B max 13.5 KG, Br 12 KG, Hc 0.8 oe; permeability, 7,000.

Stainless Steel

Material: 316L Stainless Steel + 0.5%Acrawax C

Density: 6.9 g/cc

Burn-off cycle: 425 °C (800 °F) in air for 1 h.

Sinter cycle: Evacuate vacuum chamber to less than 50 microns; heat to 815 °C (1500 °F), at maximum power; hold at 815 °C (1500 °F) for 30 min.; backfill with argon to a partial pressure of 600 microns; heat to 1290 °C (2350 °F), at a rate of 17 °C (30 °F) per min.; hold at 1290 °C (2350 °F) for 3 h; turn off power; backfill to 25 in. Hg with argon and fan-cool to room temperature.

Final properties: density, >7.35 g/cc; tensile strength, 65,000 psi; yield strength, 30,000 psi; elongation, 30%.

Cemented Carbide

Sinter cycle (see Fig 9.13): Binder removal during slow heating from room temperature to 350 °C (662 °F), with heating rate controlled to prevent rapid expansion of vapor phase from fracturing the part; heat to 600 °C (1112 °F) to complete dewaxing phase and restore vacuum to the chamber; heat rapidly to 1100 °C (2012 °F) and hold for 30 min.; heat rapidly to 1470 °C (2678 °F) and hold for 30 min.; turn off power and cool in vacuum to 1160 °C (2120 °F); backfill to near atmospheric pressure and rapid-cool in argon.

Backfill gases: Argon used early in the cycle to aid dewaxing and accelerate cooling at the end of the cycle; hydrogen used as a deoxidizer during the sinter cycle; methane used to control carbon level during sintering.

References

1) *ASM Handbook*, Vol 4, ASM International, 1991
2) *Heat Treater's Guide: Standard Practices and Procedures for Steel*, P.M. Untweiser, H.E. Boyer, and J. J. Cubbs, American Society for Metals, 1982
3) *ASM Handbook*, Vol 2, ASM International, 1990
4) *Metals Handbook*, Vol 4, 9th ed., American Society for Metals, 1981

• 10 •

Special Processes

John H. Durant, Industrial Marketing Consultant

Before and After Heat Treating

The emergence of vacuum and fully controlled atmosphere in the heat-treating industry has swept away generations of conventional wisdom in furnace charge preparation, atmosphere control, heat transfer, temperature uniformity, quality, yield, productivity, and capital requirements. The price of success in vacuum heat treating is a thorough understanding of vacuum technology summarized in Chapter 1. Information presented in Chapters 2 through 9, if heeded, will reward the user with success in his vacuum thermal processing endeavors and perhaps generate some curiosity about what else is out there. A short examination of metal processing steps prior to and subsequent to vacuum heat treating is appropriate.

Melting and Casting

Quality begins in the melt shop. As in traditional heat treating, the melter has developed techniques, equipment, procedures, and products of consistent quality, which meet the requirements of large mature markets and are cost effective. The demand for higher performance product drives the melter to employ vacuum to avoid the deleterious effects of the atmosphere on molten metal. Gases occur in metal in numerous ways: as mechanically entrapped voids; in dissolved form; and chemically as oxides, nitrides, carbides, and hydrides.

Vacuum melting furnaces were developed initially to process exotic materials such as superalloys, uranium, beryllium, molybdenum, titanium, and zirconium. As melting techniques evolved and equipment reliability and capacity increased, applications to more conventional alloys were developed. One notable example: in 1949 a 10-lb ingot of SAE 52100 steel was produced in a small laboratory vacuum induction melting furnace for a ball bearing manufacturer. Spectacular increase in fatigue life was observed and later verified on a statistically appropriate number of specimens made from the next generation of furnaces, which could produce 200-lb billets. The rest is history. This specialty steel industry entered the new market with a succession of investments that culminated in production vacuum melting furnaces capable of producing billets up to 15 tons. All of the 52100 steel produced in the USA is vacuum melted. Numerous other applications follow a similar pattern. In each instance the heat treater finds new markets. Quality produced in the melt shop must be maintained in subsequent manufacturing steps involving heat, where the surface of the metal must be protected against exposure to ambient atmosphere or imperfectly controlled furnace atmospheres. The optimum environment is vacuum or controlled-atmosphere following vacuum purging. There remain certain situations which make such idealized conditions impracticable because of operations and/or financial constraints. Nevertheless, quality, environmental, and energy efficiency concerns generate constant reminders to review the feasibility of vacuum.

The typical heat treater may be only dimly aware of the melt shop. This is not surprising, considering that a ton of steel uses only 5 ft^3 of melting furnace capacity in the crucible. Contrast this with the typical heat-treating furnace, which handles loads with a bulk density of 50 lb/ft^3: a factor of 1:10. When thin-gage assemblies are brazed, the net bulk density of product after deducting the "excess baggage" which is fixturing, may balloon this factor to 1:250. To put it another way, there is a lot of opportunity for many vacuum heat treaters to maintain the quality of the output of a few melt shops.

The choice of vacuum melting equipment is dictated first by the charge material, second by the desired form of the product, and third by the size of the melt.

The Charge Material

Ferrous alloys containing iron, nickel, or cobalt are successfully induction melted in refractory crucibles of magnesia, alumina, zirconia, where the induction field magnetically couples directly to the charge. Nonferrous alloys such as copper, aluminum, zinc, and uranium, which will not couple with the magnetic field, are melted in graphite crucibles which are susceptible to the inducted field and heat the charge by conduction and radiation.

Reactive and refractory metals, which by virtue of their high chemical activity in the molten condition, and which have very high melting points, must be handled in cold crucibles usually made of water-cooled copper.

Such an arrangement precludes the use of induction as an economical energy source, based on current technology. Instead, energy is applied directly to the charge by electric arc. The charge is fabricated into an electrode for consumable melting. Alternatively, the charge in the form of chunks may be fed to the cold crucible and melted by using a non-consumable electrode of graphite or water-cooled tungsten.

Electron beams in single and multiple array are used for the highest melting temperature materials such as niobium, tantalum, tungsten, and some superalloy applications, where the high concentration of energy and the low pressure required provide conditions which insure the highest purity of product.

Plasma arcs provide high-energy concentration and will enjoy wider use as ongoing experience establishes unique advantages and improved controls are developed.

The Form of the Product

Ingot, billet, and wire bar (*i.e.*, massive shapes for subsequent reduction to intermediate form and mill product) are produced from a pre-alloyed mixture of virgin materials, suitably prepared recycle, or scrap. Induction furnaces, usually tilt-pour in capacities of 1 ton and up, are employed. Vigorous induction stirring assists in achieving alloy uniformity. The melting cycle is adjusted to allow adequate time to pump away gaseous products which may be contained or generated by deoxidation reactions. The addition of certain alloy constituents such as niobium, titanium, and rare earths cannot be pre-alloyed owing to their extreme chemical activity with consequent risk of refractory crucible liner attack and resultant oxide contamination of the melt. Such elements are added immediately before teeming the melt, allowing only enough time to mix. Volatile alloying elements such as aluminum or manganese may be held similarly as late additions. An alternative technique is to employ a partial pressure of inert gas to suppress evaporation loss.

In numerous instances, depending on the alloy, the size of the melt, and the cross section of the ingot mold, segregation may occur during solidification. In such cases the melt is repeated, using the billet as the consumable electrode in a vacuum arc melting furnace. The incremental melting and solidification produces excellent uniformity both in chemical composition and in grain size. Vacuum arc remelted (VAR) billets command a justifiable premium price.

Reactive and refractory billets are produced by arc melting. The electrode is fabricated by welding together pieces of virgin and/or remelt stock. The resulting billet is used as the electrode in a larger VAR furnace to obtain the quality advantage of double vacuum melting.

Precision Investment Casting

A substantial share of the vacuum melting business is the production of precision investment cast shapes by highly specialized foundries which serve aerospace, chemical, petroleum, and paper making equipment as well as nuclear, medical, r & d, and weapons industries. This near-net-shape small parts production technique uses production semi-continuous VIM furnaces typically of 50 lb capacity. The system comprises a melt chamber containing the vacuum induction melting coil and crucible; a mold lock; mold elevator; the necessary control components; and two vacuum pumping systems: one for the melt chamber and one for the mold lock. An isolation valve permits the mold lock to be opened for loading and removing the mold, thereby avoiding melt chamber contamination. The crucible is charged through an interlocked vacuum billet charging mechanism. See Fig 10.1.

The system employs a water-cooled vacuum melt chamber. The induction coil is mounted inside the melt chamber so that it can be tilted to pour the melt. Surrounded by the induction coil is a crucible which contains the melt. The coil is connected to a high-frequency induction power supply. The voltage from this power supply must be low enough to eliminate any possibility of corona discharge, which may change to destructive arcing inside the vacuum chamber during the "power-on" or melting stage of the process.

Operating cycle time depends on the power supply capacity and the size of the melt. With a 75 Kw power supply, for example, a 30 lb ferrous melt will take approximately 10 min. The power supply frequency is an important consideration in vacuum induction melting. Most melting of ferrous charges in the 30 to 200 lb range is done in the 3,000 to 10,000 Hz range.

Selection of the power supply frequency is determined by the size of the melt, the basic rule being that the frequency varies inversely to the depth of penetration. As the size of the melt is reduced, for example, the frequency increases. A larger melt, e.g., 3 to 5 ft in diameter would probably require a line frequency of 60 or 180 Hz. This insures good stirring action caused by lines of electrical flux cutting through the melt.

Directional Solidification

Grain orientation control capability can be incorporated in vacuum precision investment casting. The technique uses controlled cooling of the

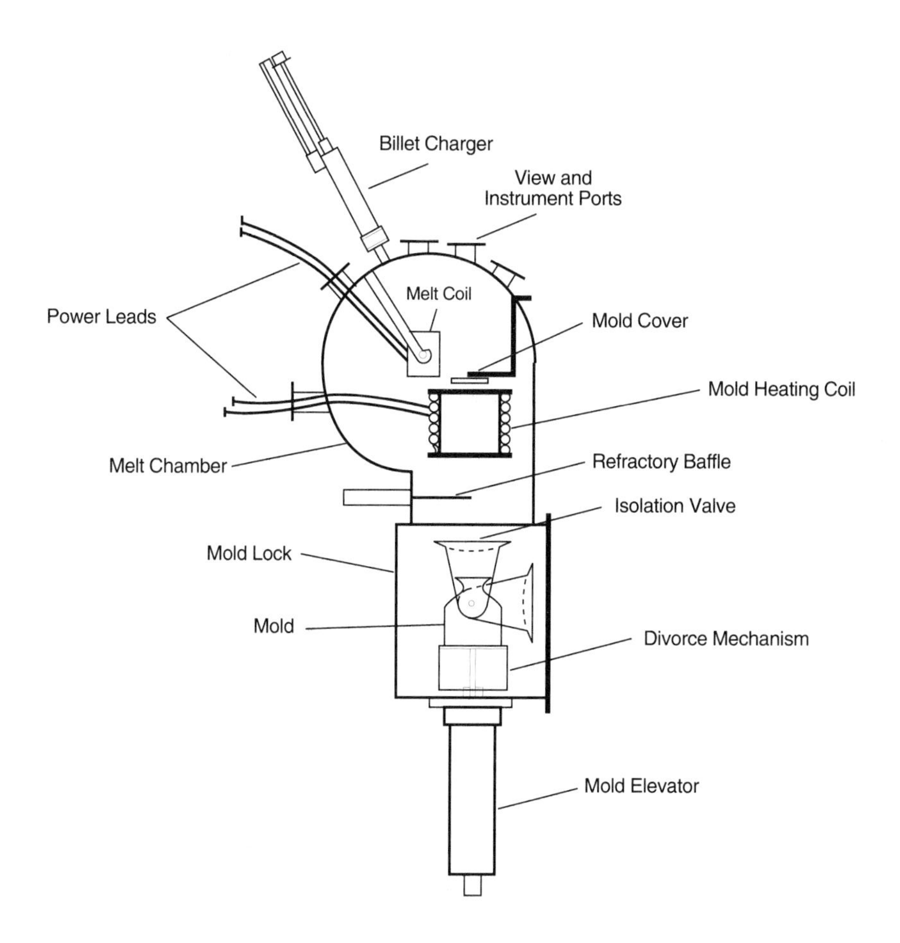

Fig 10.1 Vacuum precision investment casting furnace equipped for directional solidification. Semi-continuous operation employs the pre-alloyed billet charger at the top operating in conjunction with the mold lock at the bottom. Slight modification enables this system to produce monocrystaloy (single crystal) turbine blades.

casting to produce a high-strength, oriented-crystal part. Solidification of the material is controlled so that the casting consists of longitudinally oriented grains that result in increased axial strength, and greater creep resistance. Jet engine turbine blades are produced by this process.

Monocrystalloys

The technique has been developed even further, to produce cast blades of a single crystal which has no grain boundaries. This, in effect, eliminates additional elements required in directionally solidified and equiaxed castings to control grain boundaries. The result is a single "simplified" crystal alloy with optimized creep strength and resistance to thermal fatigue, oxidation, and hot corrosion.

Investment castings up to 500 lb are produced in VIM furnaces principally in corrosion-resistant nickel-base alloys. Some success is reported with titanium castings using refractory slurries to line graphite crucibles and skim coat graphite molds. Vacuum arc skull melting furnaces utilizing cold-copper hearths fed by consumable electrodes produce high-quality parts.

Powder Production

Alloy powders used to fabricate high-performance powder metallurgy (P/M) parts are produced in vacuum melt furnaces. Rapid solidification results when a VIM heat is atomized through a nozzle by pressurized inert gases to produce uniformly sized spherical powder. Variations of the basic technique involve pouring the molten stream on refrigerated surfaces which may be rotating drums or spinning discs. VAR is used employing high-speed rotating electrodes which throw off droplets by centrifugal force.

The Size of the Melt

The third element in the selection of vacuum melting equipment is the scale of the plant. Vacuum induction melting is the first choice when dealing with melt sizes up to 1,000 lb. Its flexibility makes it useful in a variety of moderate-production, small- to medium-size investment castings. Numerous specialty mill products, including wire, resistance ribbon, precious metal, beryllium, copper, glass-to-metal sealing alloys, and magnetic alloys are converted from vacuum-melted billets of 2,000 lb or less. A majority of the materials can be processed in refractory or graphite crucibles. Initial cost is moderate, installation costs are low, and control is relatively simple, amenable to PC computer programming.

Vacuum arc melting, when applicable to the user's process requirements, offers the lowest equipment costs. The chamber is of minimum volume, the electrode tower mechanism operates in ambient conditions where it can be easily maintained, and the power supply is the least costly. The equipment is scalable to produce large billets of 30 in. diameter and up. Installation requires pitting and elaborate instrumentation to enable remote operation for worker safety. Productivity is high and unscheduled shutdowns are rare.

Electron beam furnaces in any scale are the most costly, but they may be the obligatory choice when their capability is unique. An example is large-scale melting of alloys of niobium, tantalum, and tungsten. Hard vacuums, high voltages, and precise beam control demand sophisticated computer capability together with operators who understand their jobs unusually well.

The quality embodied in vacuum-melted cast shapes or mill product produced from vacuum-melted billets can be most successfully preserved and in many instances enhanced by good vacuum technique in subsequent operations involving thermal processing, be it heat treating, annealing, sintering, joining, coating, or surface enhancement. The heat treater must insist on receiving reliable information on prior processing history of the material he accepts, if he is expected to produce optimum quality output.

Metal Injection Molding

Metal injection molding (MIM) is the term applied to the production of small, intricate metal parts by a combination of injection molding and sintering. Fine, spherical metal powder is intimately mixed with a plastic binder, usually in the ratio 55 to 60 vol% metal powder to 40 to 45 vol% plastic. The mixture is placed in an injection molding machine, heated to the flow point of the plastic, and forced into a mold under pressure. When the mix has solidified in the water-cooled mold, the mold opens and the part is removed. This part of the process is nearly identical to the procedure used to make the enormous variety of common and inexpensive thermoplastic articles. In the case of MIM, however, the process is not yet complete. After molding, the plastic binder must be carefully removed, allowing the metal grains to touch each other. The part is then sintered further to consolidate the grains, thereby producing a strong, useful metal article. Markets include watch cases, automotive parts, firearms, business machines, sewing machine parts, and orthodontics, to name a few.

MIM is used to produce quantities of precision parts whose shapes are more complex and therefore not conducive to production by the normal P/M method of pressing metal powder followed by sintering. As a process, therefore, it competes with precision investment casting and machining from the solid, or with mechanical assembly of several parts. The MIM process is also predictable because the green part shrinks linearly upon binder removal and sintering.

Shrinkage is about 20%, depending on the exact mixture used, and is precise from part to part. The standard part-to-part dimensional tolerance is ±0.003 in./in. The part surface is smooth and fine, thereby eliminating secondary machining operations in most cases.

Binder Removal

The fugitive binder is removed by thermal decomposition in the sintering furnace. This technique is preferred because of the high losses experienced in transferring the extremely fragile, binderless parts from a separate debinder system to the sinter furnace. Since it is impractical to move the debindered parts any distance, the commercial heat treater can expect to receive either pre-sintered parts, or more likely green parts fresh from the mold, for final thermal processing. The first step is binder removal, followed by sintering without moving the workpieces in the furnace.

The most popular binder in use today is a two-part mixture of ordinary paraffin wax and a thermoplastic such as polypropylene. With small amounts of wetting agents and surfactants added, the material molds well. In thermal debinding, the paraffin is removed first, followed by the polymer, based on their relative vapor pressures. The part shrinks slightly as the

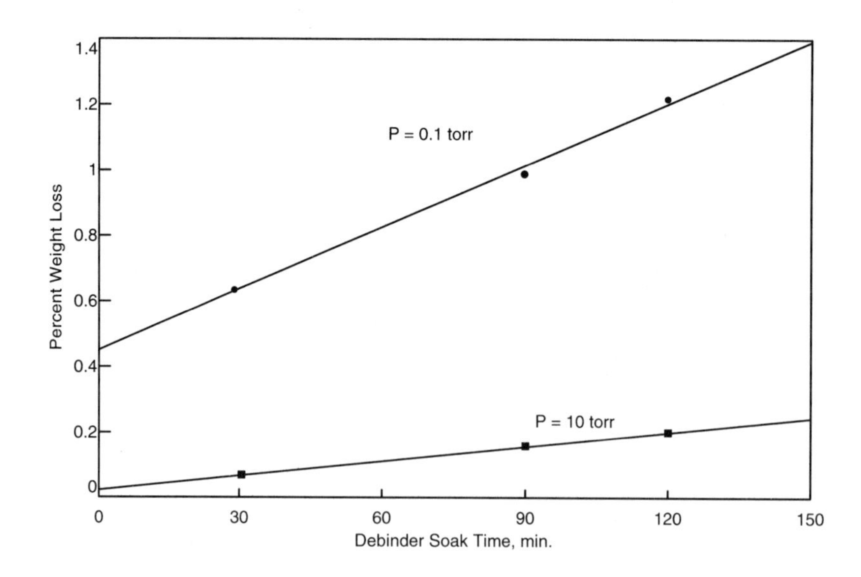

Fig 10.2 Effect of pressure on debinder time

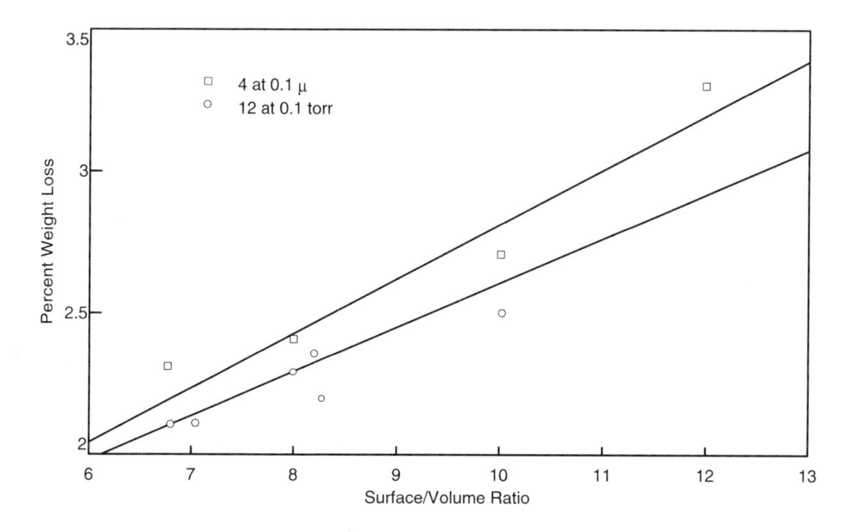

Fig 10.3 Effect of shape factor on debinding

paraffin is removed, leaving support dependant on the polymer, which breaks down and is removed at higher temperature. A debinder schedule can be determined after the exact nature of the binder is established.

Advantage of Vacuum

The debinder schedule for a typical part with a paraffin-polymer binder in air atmosphere requires 24 to 48 h. The wax and polymer are burned slowly and the metal grains heavily oxidized. In vacuum, the rate of removal is faster because the molecules evaporate from the surface after they pass through the spaces between the grains in the matrix. In air, evaporation from the surface is suppressed. The effect of low pressure is shown in Figs 10.2 and 10.3. The wax molecules are relatively large in relation to the spaces between the grains through which they must pass, and flow is therefore limited. If the furnace temperature is raised too high, the vapor pressure of the wax increases because of the constraint of the intergranular passages, and the part may blow up. Also, some geometries will cause the part to slump and distort if the paraffin is melted. Successful debinding depends on sublimating the paraffin by controlling the part temperature closely at very low temperatures. A ramp from room temperature to 110 °C (230 °F) followed by a soak at a pressure level below 1×10^{-3} torr will succeed. Carbon

20 wax, the most common analysis, has a vapor pressure of only about 150 microns at this temperature. Wax vapor pressures are shown in Fig 10.4.

The polymer constituent of the binder does not evaporate until it undergoes thermal decomposition at about 400 °C (750 °F). It breaks into a wide assortment of compounds of relatively high vapor pressures, which can be removed in a stream of flowing inert gas at temperatures above 400 °C (750 °F).

This distinction is important because some polymer breakdown products condense in the diffusion pump, necessitating frequent fluid replacement. The paraffin, on the other hand, remains in vapor form at diffusion pump boiler pressures and temperatures and therefore passes through. Thus, in vacuum debinding parts with paraffin-polymer binders, the paraffin is removed at low temperatures by sublimation at diffusion pump pressures; the polymer at high temperatures and at mechanical-pump pressures. Figure 10.5 illustrates a typical cycle.

The spent paraffin binder condenses on the first cold surface it encounters, and the cold-wall chamber soon acquires a heavy layer of recondensed binder, which necessitates frequent cleaning. This difficulty is overcome by operating the chamber with water heated to 55 to 60 °C (130 to 140 °F) during debinding operations. At this temperature the binder is fluid and runs down

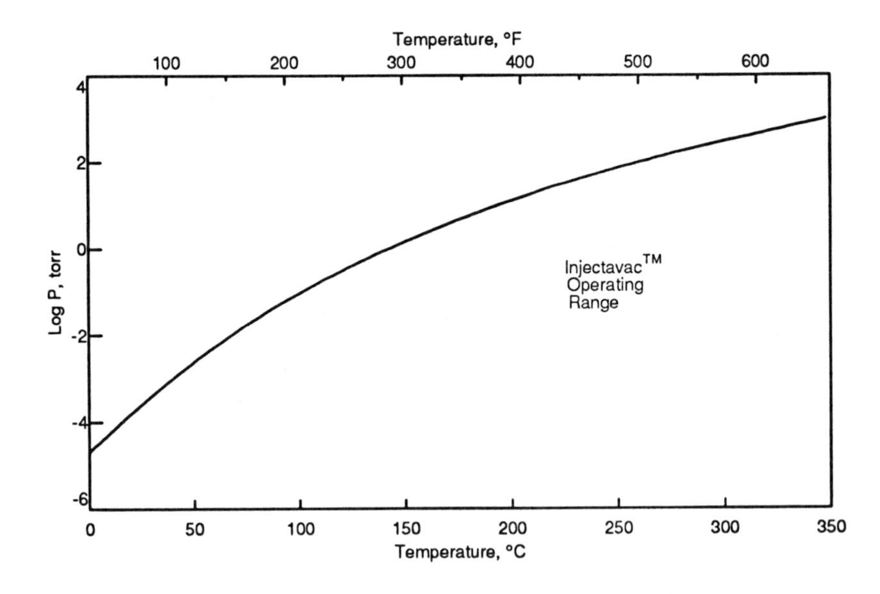

Fig 10.4 Vapor pressure of Carbon 20 paraffin wax

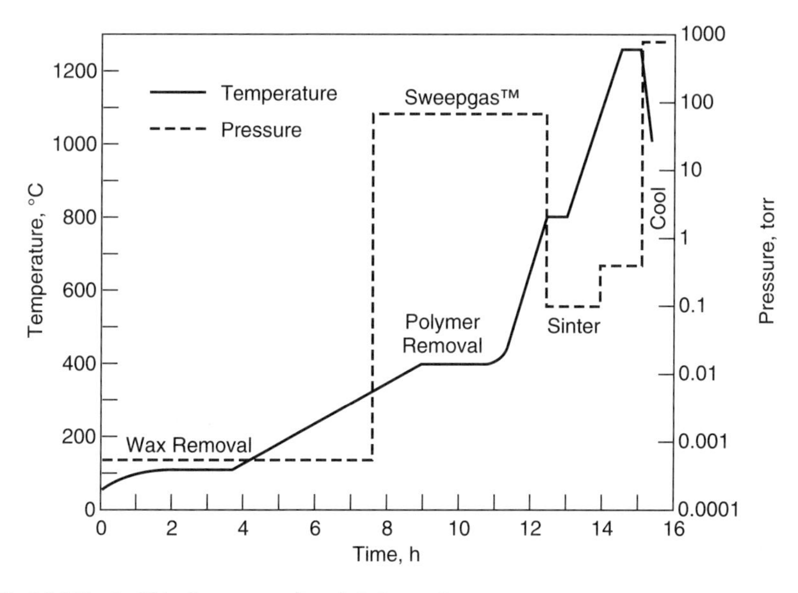

Fig 10.5 Typical binder removal and sinter cycle

to a reservoir where it may be collected. The remaining thin film is negligible. Condensation also occurs in the pumping manifolds, unless they are kept warm. A special condenser is required to prevent the vapor from entering the mechanical pump. Unprotected standard pumps choke and seize on the accumulated condensed vapor. The condenser should present a very large condensing surface and retain good conductance after all vapor has been removed in each run. It should be designed for rapid stripping between runs. Figure 10.6 illustrates a typical condenser.

A more elegant solution to the problem of spent binder condensation in mechanical pumps is the use of a pump that is self-protecting without need for a condenser. In such pumps a small amount of sealing and lubricating oil is injected at each rotation. The oil entrains the spent binder products and is ejected from the pump to a reservoir. The spent oil is resold to reclaimers or incinerated in boilers, thus keeping oil cost at a minimum.

Carbon Control

Early development established that carbon control in finished parts can be optimized by avoiding long storage times between molding and sintering. The reduction in average carbon level is caused by excess oxygen from copolymers in the binder reacting with carbon contained in the tiny (less than 10 micrometers diameter) metal grains. The grain surfaces become oxidized by the binder but are reduced again at sinter temperatures by carbon diffusing to the surface oxide. The oxide can be reduced by hydrogen at a temperature of 400 °C (750 °F). This temperature is lower than that needed for rapid reaction with carbon. Retention of nearly all the carbon can be achieved by introducing small quantities of hydrogen at 400 °C (750 °F). The scientific foundation for this technology has been established by Finn and Thompson.

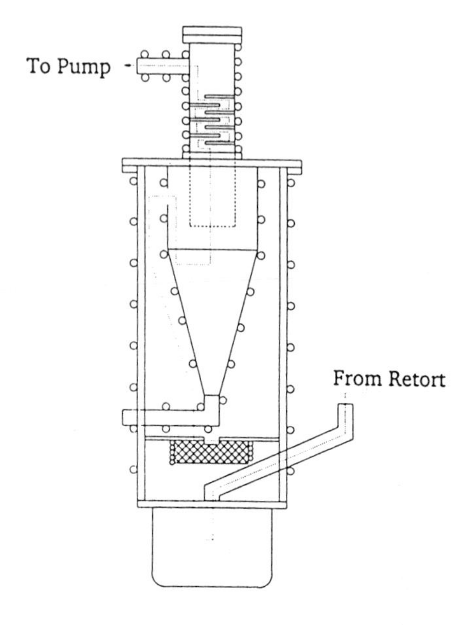

Fig 10.6 Sweepgas condenser detail

In operation, the furnace pressure is raised to 10 to 20 torr by the introduction of hydrogen. This pressure is well below the 30 torr level at which a partial pressure of hydrogen in air can explode. After a short 1 to 3 min. soak in the gas atmosphere at 400 °C (750 °F), the hydrogen is pumped out. The process is repeated several times. It is assumed that the gas fills the pores of the metal matrix, reacts with the oxides on the grain surfaces and is pumped out as water vapor. All oxygen is removed by repeated cycles. The result is production of parts with only very slight (0.05%) reduction in carbon content.

Sintering

Most MIM parts are required to have a density of 97 to 98% of theoretical to achieve the strength needed in their application. These high densities are made possible by using extremely fine spherical powders and by sintering at unusually high temperature. Iron-silicon alloys and nickel-iron alloys usually reach high density after a 1 h soak at 1290 to 1315 °C (2350 to 2400 °F). The austenitic stainless steels such as Type 316, often specified for orthodontic braces and medical instruments, achieve highest density by sintering at 1350 °C (2460 °F).

At these elevated temperatures some alloying elements, notably chromium and manganese, will evaporate at the pressures provided by the diffusion pump. A backfill of inert gas is necessary to suppress the alloy vapor and the loss of constituents.

Conclusion

Specialized vacuum furnaces are required to realize optimum quality and productivity in a single combination debinder-sinter cycle. The power supply must be capable of being turned down to allow controlled heating below 90 °C (200 °F) for paraffin removal; a hot-water recirculation system keeps the chamber warm enough to condense the spent binder vapor, yet maintain it in a fluid state for collection in a removable reservoir; the mechanical pump is protected by a vapor condenser or is the once-through lubricant type; and a low-pressure hydrogen system assures carbon control.

These capabilities and facilities are seldom, if ever, found in standard vacuum heat-treating furnaces. The usual heat-treating furnace, however, is quite capable of removing the last traces of binder and of sintering the parts. The binder may be removed by heating the workpieces slowly in an oven with a flow of inert gas such as nitrogen to carry away the spent binder products without allowing the workpieces to be excessively oxidized. In this manner, parts can be successfully produced by a two-step process which is, however, much less efficient than the single-step process performed in the specialized furnaces. Just enough binder is left after the debinding step to allow the parts to be handled gingerly and placed in the vacuum furnace for completion. By closely controlling the amount of binder left in the parts, fouling of the vacuum furnace can be limited.

Surface Enhancement

The heat treater is familiar with traditional methods of modifying surfaces of steels primarily to enhance hardness to varying degrees and to various depths. The use of conventional atmosphere furnaces in conjunction with the introduction of hardening agents such as carbon, nitrogen, and other elements in gaseous, liquid, or solid form is well established and widely practiced.

The arrival of vacuum as an accepted industrial processing tool presents a broad spectrum of new possibilities for surface enhancement. The origin of a large amount of this new technology is the research and development which fueled rapid progress in microcircuits involving controlled deposition (as well as removal) of conductors, insulators, and semiconductors. The heat treater's interest lies in the trickle-down technology which can be utilized in industrial parts processing. Surface enhancement processes of primary interest together with their principal characteristics are summarized in Table 10.1.

Benefits include hardening, wear enhancement, reduced friction, corrosion resistance, bonding layers, thermal and electrical insulation, thermal and electrical conductivity, control of optical and thermal transmission (filters) and reflectance (mirrors), attractive appearance.

The root technologies from which numerous hybrids evolved include those discussed below.

Chemical Vapor Deposition

Chemical Vapor Deposition (CVD) is the process of reacting gaseous constituents to form a film on a surface. The starting material or precursor is a compound of the material to be deposited. Thin films, thick films, or monolithic structures may be deposited. Various reactions may be involved.

Table 10.1 Vacuum processes used in surface enhancement

Process Characteristic	Thermal Evaporation	DC Sputtering	RF Sputtering	Chemical Vapor Deposition	Ion Implantation
COATING					
Metal	X	X		X	X
Ceramic			X	X	
Polymer		X	X		
APPLICATION					
Optical	X	X	X		
Decorative	X	X			
Microcircuit fabrication	X	X	X	X	X
Corrosion resistance	X	X			
Friction resistance	X	X			X
Surface modification					X
COST					
High					X
Moderate		X	X	X	
Low	X				
VACUUM					
Hard <0.1 torr	X	X	X		X
Soft >0.1 torr				X	
Hazardous effluents	No	No	No	Yes	Yes

Pyrolysis. Pyrolysis (or thermal decomposition) is the simplest, involving the injection of the precursor into the reactor, where it decomposes to form a film while the byproduct gas is pumped away. For example, if methane (CH_4) is injected into a thermal reactor it decomposes to form a carbon film. At high temperatures up to 2100 °C (3920 °F), a graphite film is formed which may be adjusted by varying conditions to achieve special properties. A typical product is pyrolytic graphite, which has a useful anisotropic crystalline structure that causes it to conduct heat preferentially in one direction. Automotive cylinder-head gaskets of pyrolytic graphite have enabled manufacturers to extend engine warranty periods and discontinue the use of toxic asbestos.

Thermal Reduction. Thermal reduction is a more commonly used CVD process employing a precursor, typically a halide in gaseous or vapor form, which is injected into the thermal reactor together with a reducing gas such as hydrogen or ammonia. Thermal reduction occurs on the substrate, thereby producing dense, pure, void-free films of metals, ceramics, or combinations (cermets). Temperatures range from 200 to 1400 °C (390 to 2550 °F). Operating pressures between 0.5 and 60 torr with flowing gas are typical after the system has been initially outgassed at 0.05 torr.

Materials Used. CVD metal films include aluminum, copper, gold, molybdenum, platinum, and tungsten. Ceramics which may be deposited include aluminum oxide, boron nitride, hafnium carbide, hafnium nitride, silicon carbide, silicon nitride, titanium carbide, and titanium diboride.

Structure of CVD Applications. Four different types of products have been made via CVD: fibers, powders, thin film coatings, and thick monolithic structures. Applications for the most part are exotic, high technology, high cost items which will be of interest to the heat treater as future trickle-down technology reaches industrial application.

The Furnace (Thermal Reactor). Hot-Wall Furnace designs are employed using muffles of alloy suitable for the temperature and corrosion conditions encountered at temperatures below 1000 °C (1830 °F). Heating is by an external electric furnace. Quartz muffles with heating by induction are employed where corrosion conditions demand it and the scale of the operation is sufficiently small to enable the degree of care and delicate handling required to avoid breakage.

Higher reaction temperatures require cold-wall furnace technology most generally utilizing graphite hot zones. If the CVD reaction involves a product which can damage the heating element and/or impair the efficiency of the insulation, a graphite retort muffle with vacuum pump-out and gas injection provisions is utilized in conjunction with inert gas blanketing outside the retort to insure that the reactants are confined (Fig 10.7).

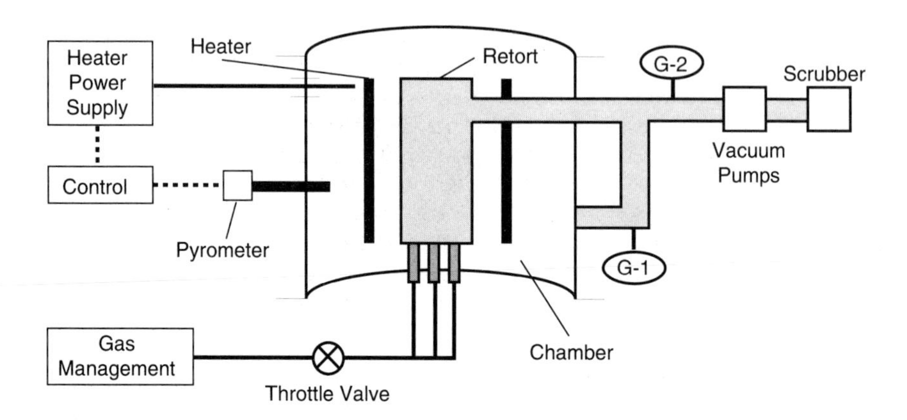

Fig 10.7 High-temperature CVD system

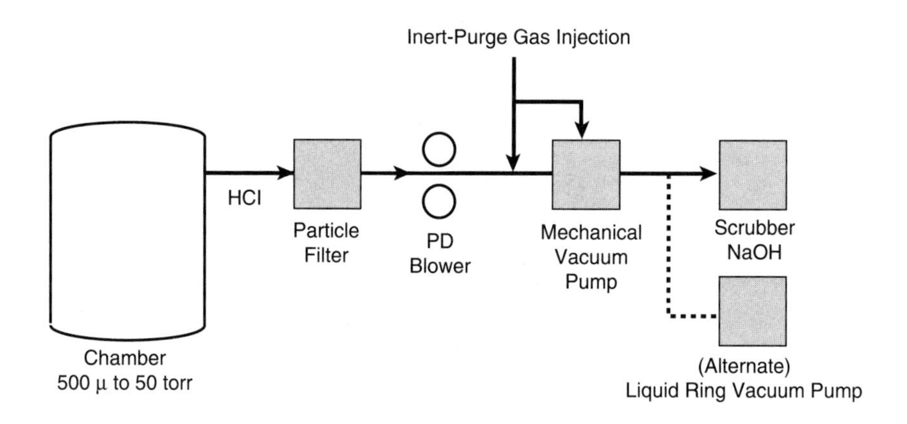

Fig 10.8 Furnace pumping system with by-product provisions

Environmental concerns are among the barriers to the acceptance of CVD. Although the reaction is conducted under vacuum, thus confining the process, there remains the problem of capturing and disposing of byproduct which is basically hydrochloric acid after the HCl off-gas comes in contact with moisture contained in the air. A rigorous inspection program must be closely managed for monitoring key reactor components for corrosion damage and replacing them on a failure-avoidance basis (Fig 10.8).

Physical Vapor Deposition

Physical Vapor Deposition (PVD) is based on evaporation and sputtering, the grandfather vacuum deposition phenomena first observed a century ago. Exploitation as economical, high-volume, large-scale industrial processes followed the emergence of vacuum systems from the component incandescent lamp and vacuum tube manufacturing status of the 1930's to the present.

Evaporation. Evaporated films are produced by heating a metallic element in vacuum. The molecules of vapor are thermally accelerated from the hot source in a straight line to any cool surface in the system. Reference to the table of vapor pressure for the elements in Chapter 1 will reveal that vacuum in the range of 1×10^{-1} to 1×10^{-8} torr enables evaporation or sublimation of virtually any element. Optical, microcircuit, electrical, decorative, vapor barrier, and wear-resistance applications are numerous and widespread.

It is obvious that such an extensively used process must offer utility, functional results, low utility and materials costs, operator safety, minimum environmental concerns, and reasonable capital costs. Users have applied its advantages and circumvented its limitations. Science and technology have developed, and industry has brought plasma-assisted evaporation to market.

Plasma-Enhanced Evaporation. Vacuum thermal evaporation produces thin films at moderately slow rates. The impact energy of the molecule of evaporant arriving at the substrate or surface to be coated is dependent on the thermal energy differential between the source and the substrate, thus limiting adhesion and film structure possibilities. Thermal treatment to diffusion-bond the molecules deposited into the substrate results in the virtual disappearance of the film, which is only a few molecules thick, as it diffuses beneath the surface.

The addition of a 2000 vdc plasma to the basic thermal evaporation process broadens application possibilities. The substrate surface is first scoured by bombardment with ionized inert gas molecules to insure the best possible adhesion. The voltage differential enhances the impact energy of the evaporant on the substrate and accelerates the rate of deposition. The source can be continuously replenished to produce thick coatings.

The aerospace industry employs plasma-enhanced evaporation (frequently referred to as "ion vapor deposition") to the production of thick, dense, adherent films for corrosion protection of aluminum airframe sections, titanium fasteners, highly stressed steel landing gear components, and bolts. Corrosion resistance exceeds electroplated cadmium and is effective at temperatures to 495 °C (925 °F). Steel parts benefit by no hydrogen involvement, and avoid embrittlement. The clean PVD process eliminates problems of hazardous waste disposal which is involved with chemical cleaning and electroplating.

Sputtering. Sputtering is a coating process which transfers material from a target material to be deposited on a substrate by bombarding the negatively charged target under a partial vacuum with ionized inert gas (usually argon) molecules. The molecule of target material thus eroded is accelerated to the substrate by a suitable voltage differential (Fig 10.9). Direct current is employed to sputter deposit metals. Radio frequency alternating current is used to sputter deposit nonconducting materials, including ceramics and polymers. Temperatures involved are low, allowing heat-sensitive targets and substrates to be processed. Impact energy is high, insuring tenacious coatings. Sputtering rates are low, on the order of a few hundred angstrom units per minute. Rates can be enhanced by arranging magnets within the target backup electrode to boost rates by a factor of 10 or more. Magnetically enhanced or "magnetron" sputtering is used almost exclusively in industrial applications.

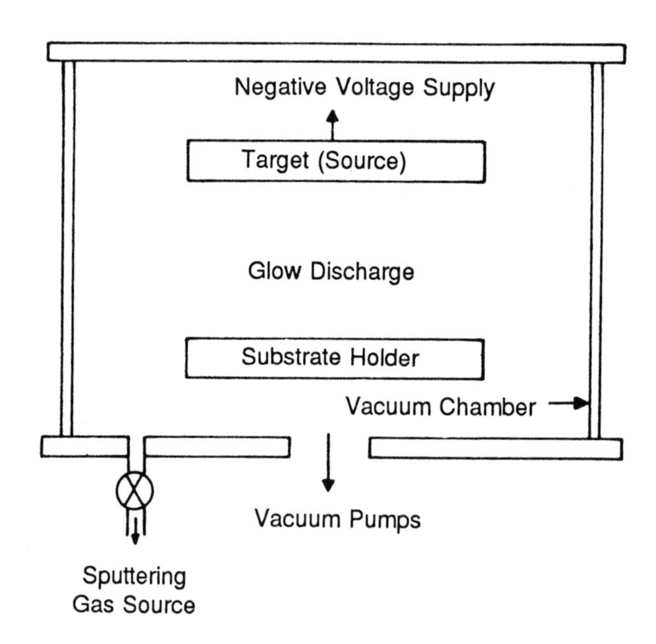

Fig 10.9 Simplified cross section of a sputtering system

PVD processes require higher and cleaner vacuums than either heat treating or vacuum melting. Stainless steel chamber construction is prevalent. Vacuums in the range of 10^{-6} torr and below require diffusion, turbomolecular, or cryogenic pumps. Diffusion pumped systems are usually trapped to exclude oil molecules from the coating chamber.

Ion Implantation. Ion implantation is an emerging PVD process which is well established in a few critical, ultra precision applications and shows promise for wider use in broader-based, more mundane uses. Unlike the surface layer coating of evaporation and sputtering, ion implantation injects molecules which penetrate beneath the surface of the substrate to produce modifications which are exempt from delamination, do not alter dimensions, and can produce an alloy gradient of precisely controlled composition and depth.

In the ion implantation process, energetic ions are accelerated and made to strike the surfaces of workpieces in the vacuum chamber (Fig 10.10). The

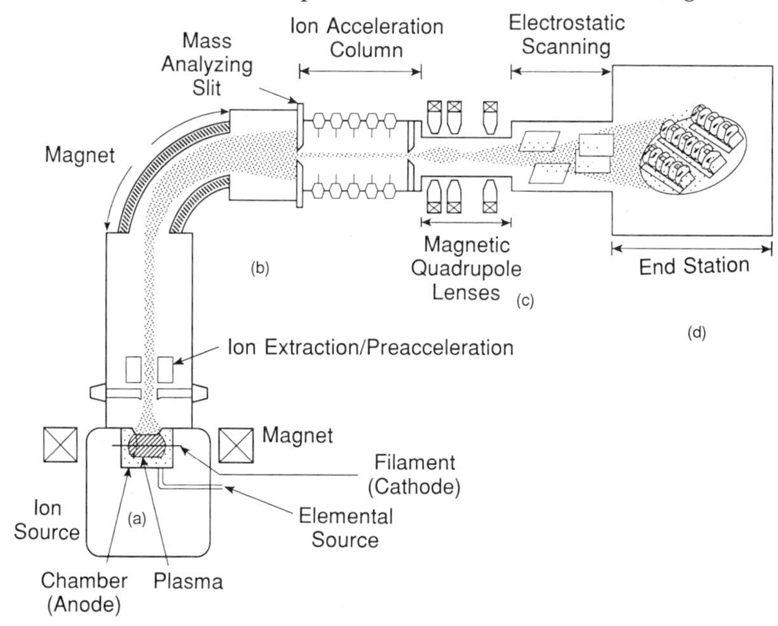

Fig 10.10 Ion implantation system. The ion implantation system comprises: **(a)** Positive ion source utilizing an elemental coating substance in either chemical or physical vapor form. **(b)** Ion separation and acceleration stages. **(c)** Ion beam focussing and scanning. **(d)** End station where the fixtured workpieces are placed for optimum exposure to the impinging beam. Illustration courtesy Spire Corporation

ions, typically with energies of hundreds of kilo electron-volts (keV) penetrate hundreds of atomic layers beneath the surface, where they are slowed and eventually stopped, through collisions with atoms of the host material or substrate. The collision cascade of host atoms creates a region of extensive "radiation damage" within the surface layer of the substrate. (Radiation damage refers to alterations produced in the crystal structure of a material.) In ion implantation of metal workpieces these alterations are extremely desirable because along with the foreign implanted ions the desired altered surface properties are obtained. For example, the combination can create an amorphous layer with no grain boundaries, ideal for providing superior wear performance. In ferrous alloys such an amorphous surface can avoid corrosion which initiates at grain boundaries.

Process limitations which restrict ion implantation to a relatively few cutting-edge technologies include:

1) Material to be coated subsurface must be in a gaseous form
2) Ion beam focussing, separation, and acceleration require elaborate magnetic, electrostatic, and high-voltage provisions which are alien to most factory environments
3) Capital costs are high, particularly in relation to production rates and workpiece size

Despite its limitations, ion implantation has captured important niche markets in orthopedic and dental implants, instrument bearings, critical cutting tools, and nuclear reactor components.

Ion-Beam-Assisted Deposition. Another hybrid, ion-beam-assisted deposition (IBAD) produces surface layers similar to those achieved by sputtering and evaporation, but which have superior adhesion and composition gradients at the interface, by virtue of high impact energy achieved by simultaneous ion beam bombardment from a 2-keV source. A typical system in commercial use employs a multiple pocket 180° electron beam evaporator which can generate a choice of metallic or metalloid vapors. Two ion sources, one high-energy for enhancing the deposition, the other of lower energy for cleaning the substrate prior to coating, are arranged to process substrates on a controlled temperature rotating holder. The resulting dual-ion-beam-assisted deposition (DIBAD) system combines desirable features of evaporation and ion implantation (see Fig 10.11).

Coatings can be grown from within the substrate, for best adhesion. Ion beams ensure high density, pin-hole-free film, and can be controlled for desired porosity. High through-put, and low-cost production, as compared with ion implantation, is possible.

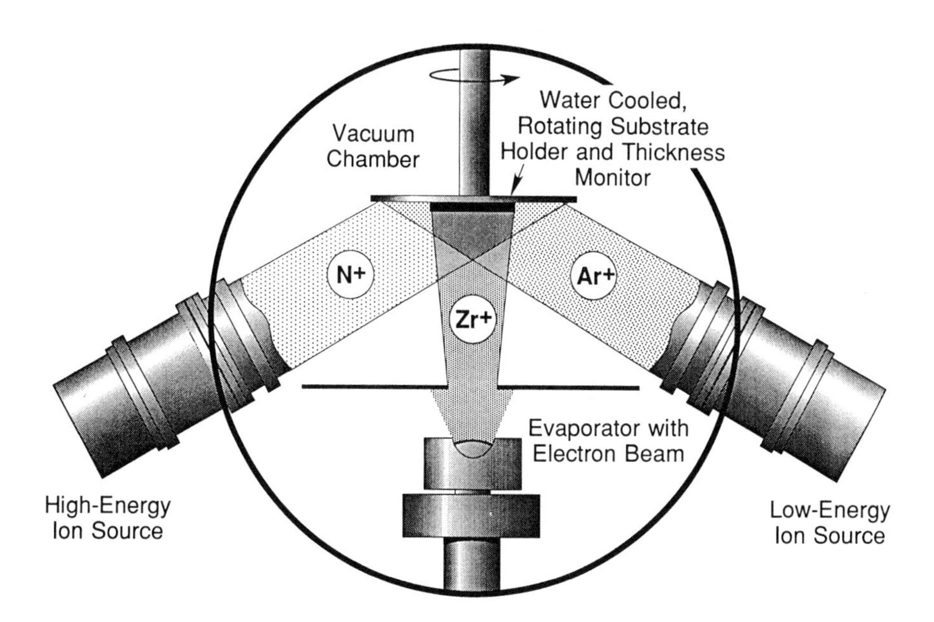

Fig 10.11 Dual-Ion-Beam-Assisted Deposition System. This system employs an evaporator with electron beam to produce metal molecules while a high-energy ion source supplies accelerated (5 to 40 keV) gas molecules to form a compound on the substrate. The low-energy ion source is used initially to bombard the substrate at 2 keV with argon for cleaning and/or to modify the surface texture. Illustration courtesy Spire Corporation

Concluding Remarks and Acknowledgements

It is hoped that this sampling of upstream and downstream vacuum processes provides the heat treater with a wider vision of how his services interface in terms of materials, properties, quality objectives, and market possibilities.

The author extends sincere expressions of appreciation to the many co-workers, suppliers and customers who collectively enriched his 30 years of association at Vacuum Industries. The knowledge thus accumulated made it possible to undertake this overview chapter. Particular thanks go to K.W. Doak for his counsel in planning and outlining; to S.W. Kennedy for contributing the section on Metal Injection Molding; and to Veronica Flint of the ASM International Reference Publications Department for her patient persistence and encouragement which brought this material to press.

References

1) M.J. Blasko, K.W. Doak, and P.H. Hanson, "Trends in Vacuum Precision Casting," *Vacuum Metallurgy Conference on Specialty Metals Melting and Processing*, ISS, 1986, pp 27-31
2) R.M. German, "Variations in the Essential Powder Injection Molding Processes," *Business Opportunities in the Emerging Markets for Injection Molded Metal, Ceramic and Cemented Carbide Parts*, Gorham Advanced Materials Institute, Monterey, CA, 1989
3) S.W. Kennedy and C.W. Finn, "MIM Part Debinding and Sintering in Vacuum Furnaces," *Business Opportunities in the Emerging Markets for Injection Molded Metal, Ceramic and Cemented Carbide Parts*, Gorham Advanced Materials Institute, Monterey, CA, 1989
4) C.W. Finn, Vacuum Binder Removal and Collection, *International Journal of Powder Metallury*, Vol 27, No. 2, 1991
5) C.W. Finn and E.D. Thompson, *Carbon Control during Debinding and Sintering of Metal Injection Molded Steel Parts*, The Metals Society, 1990
6) R.M. German, *Powder Injection Molding*, Metal Powder Industries, Princeton, NJ
7) *Metals Handbook*, 9th ed., Vol 7, American Society for Metals, 1984
8) *Thin Film Processes*, John L. Vossen, Werner Kern, Eds., Academic Press Inc., 1978, pp 257, 335, 401
9) Paul J. Timmel, "The Role of CVD in Ceramics Processing," presented at 39th Pacific Coast Regional Meeting of The American Ceramic Society, Seattle WA, October 1986
10) D.P. Stinton, T.M. Besmann, and R.A. Lowde, *Advanced Ceramics by Chemical Vapor Deposition Techniques*, Oak Ridge National Laboratory Ceramic Bulletin, Vol. 67, No. 2, 1988
11) J.L. Vossen et al., *Pumping Hazardous Gases*, American Vacuum Society Transactions, 1980 Symposium
12) P. Sioshansi, *Nuclear Instrumentation Methods*, B 24/25, 1987, p 767-7700
13) B. Haywood, *Advanced Materials and Processes*, Vol. 138, December 1990
14) V.L. Holmes, D.E. Muelberger, and J.J. Reilly, *The Substitution of IVD Aluminum for Cadmium*, McDonnell Aircraft Co., St. Louis, MO
15) G. Legg, *Ion Vapor Deposited Coatings for Improved Corrosion Protection*, Abar Ipsen Industries, Bensalem PA 19020

• 11 •

Inspection and Quality Control

Jeffrey A. Conybear, Metal-Lab, Inc.

Introduction

In discussing quality control and inspection of vacuum systems, the first step is to define not only what is meant by vacuum but also the behavior of any element within it. The heat treat shop, manufacturing facility, and laboratory need to know the level of vacuum that is needed and acceptable for their particular use and situation.

What Is a Vacuum?

Vacuum is derived from the Greek word vacuus meaning empty. In vaccum, the object is to achieve an empty chamber, or, more precisely, a chamber that is devoid of any particles, vapors, gases, or matter. If, in fact, this were possible the chamber would also be devoid of pressure. However, the equipment that is available today is capable of removing enough atoms from the vacuum chamber to enable the creation of a contamination-free environment.

Achievement of vacuum is expressed in terms of pressure relative to a standard pressure. Standard pressure is the pressure exerted by the atmosphere under normal conditions at sea level, and is equivalent to approximately 14.7 lbs per square inch. In a well-type manometer, this is also the equivalent pressure needed to support a column of mercury to a height of 30 in. In terms of vacuum measurement the following pressure relations are normally used:

One standard atmosphere = 29.921 in. Hg = 14.696 lbs per sq. in.

One standard atmosphere = 760 mmHg = 760,000 microns Hg = 760 torr

Figure 11-1 illustrates these relationships.

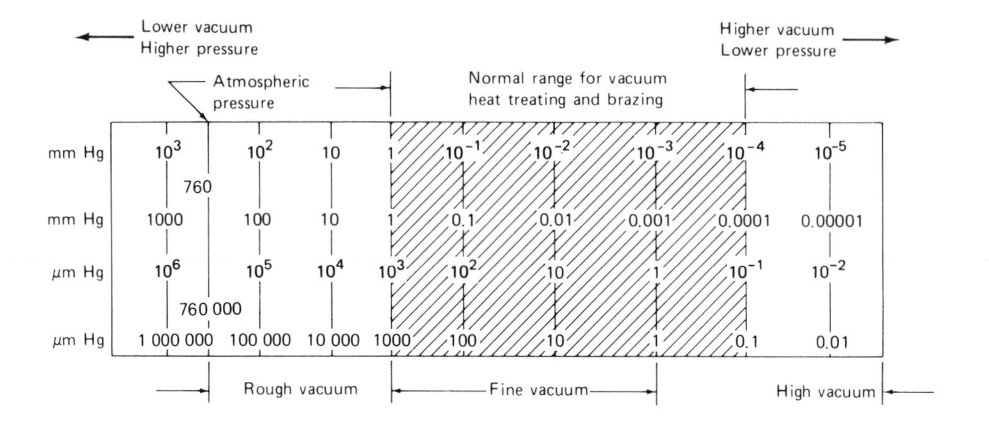

Fig 11-1 Range of pressure and vacuum

Gases

Before pumpdown in a vacuum chamber, the constant motion of gas molecules causes collisions between the molecules and the chamber walls. These collisions exert a force of one atmosphere on the chamber. The number of molecules decreases as the air is pumped out of the vacuum chamber. This causes a decrease in pressure. When investigating the causes of possible leaks and pumpdown performance in a system, an understanding of the behavior of gases in a vacuum is particularly important.

The behavior of gases under pressure illustrates the importance of having clean, dry work before vacuum processing. When troubleshooting a system, during a pumpdown, or when leak detecting, knowledge of the behavior of water vapor and other gases can be useful. Figures 11-2 to 11-5 and Tables 11-1 and 11-2 present information for certain gases under various pressures and temperatures.

In a conventional vacuum heat treating process, as the system is being pumped down the gas molecules in the chamber are evacuated to the outside through a combination of pumps on the furnace. As more and more air is pumped out of the furnace, the Mean Free Path (MFP) of the molecules in the furnace changes. The Mean Free Path is described as the "average

Table 11.1 Pressure of Gases in Air in Relation to Their Volume Percent

Gas	Volume Percent	Partial Pressure in Torr
Nitrogen	78.08	593.4
Oxygen	20.95	159.2
Argon	0.93	7.1
Carbon Dioxide	0.03	0.25
Neon	0.0018	0.00138
Helium	0.0005	0.0004
Krypton	0.0001	0.0000866
Hydrogen	0.00005	0.000038
Zenon	0.0000087	0.0000066

Table 11-2 Vapor Pressure of Water at Various Temperatures

Temperature °F	Pressure in Torr
212	760 (boiling)
122	98
77	17
–104	0.1
–173	5×10^4

distance molecules will travel before colliding." At atmospheric pressure the MFP is very short; but it increases dramatically under vacuum. Under atmosphere to about 10^{-3} torr the pressure is high enough that the MFP is relatively short, and the movement of atoms and molecules is similar to the movement of billiard balls, and is described as being in the "viscous flow range." However, at lower pressure (or high vacuum, 10^{-3} torr) the MFP increases and the movement of the molecules becomes more unpredictable. In this range, the flow of molecules is known as "molecular flow." The molecules in the chamber will stick to the walls for a short period of time and then go off in a random direction, migrating to the pumps only on their own accord and the mechanical pumps are unable to evacuate the chamber further.

In the molecular flow range the movement of the molecules is described in terms of conductance. A one square inch opening has a conductance of about 781/s and the molecules will only move into the pumps at random. Conductance must be improved in order to make it easier for the molecules to find the pump and effectively evacuate more molecules. When moving molecules out of the system the size of the opening to the pumps is the limiting factor—not the size of the pump.

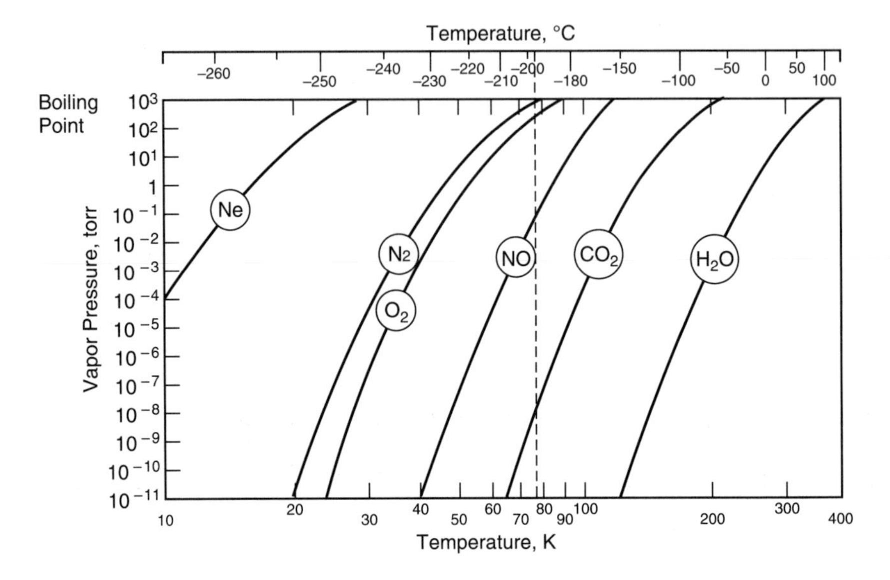

Fig 11-2 Vapor pressure curves of various gases. Curves to the right of line are low vapor pressure, to the left high vapor pressure. Dotted line is the boiling point of liquid nitrogen, –196 °C

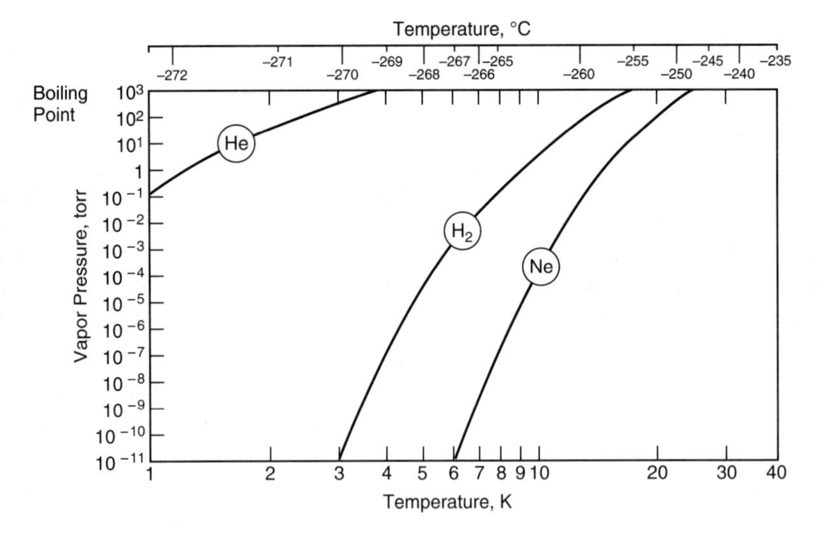

Fig 11-3 Vapor pressure of various gases. High vapor pressure at extremely low temperatures

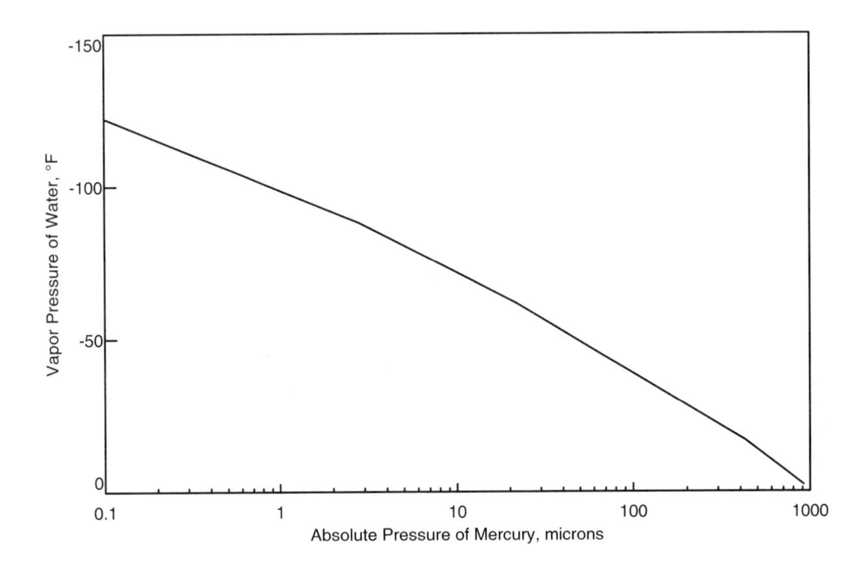

Fig 11-4 Vapor pressure of water in equilibrium with vacuum

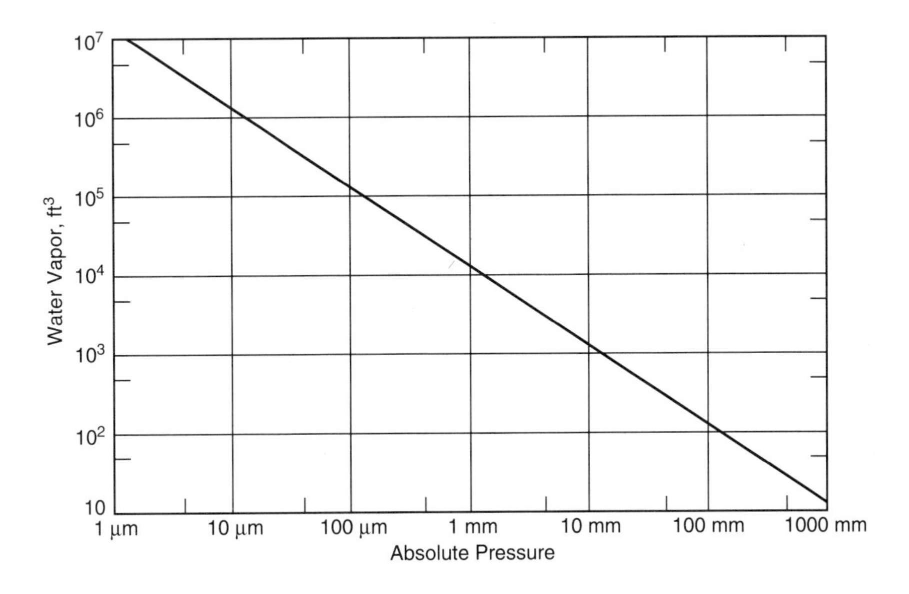

Fig 11-5 Volume of 1 lb. of water at 20 °C under a vacuum

When discussing the Through-put of a system, the behavior of gases under low pressure is important as it impacts upon the time it takes to pumpdown to low pressure in the range of 10^{-6} and lower.

Gas Laws

The volume that a gas occupies is related to pressure. This relationship is described simply by Boyle's Law: $V_1P_1 = V_2P_2$, where temperature is held constant. As indicated by Boyle's Law, under a vacuum the volume occupied by a gas increases significantly as the pressure if lowered. This is readily demonstrated in the time it takes to pumpdown a system or maintain pressure when dirty or greasy parts are in the furnace. The gas absorbed in grease evolves or outgasses slowly, and until all outgassing is complete a satisfactory vacuum cannot be achieved.

If temperature is added to the equation, the same effects can be seen. Known as the General Gas Law: $P_1V_1/T_1 = P_2V_2/T_2$, temperature is expressed in Degrees Kelvin. Together with Charles' Law (the proportion of ratio between volume and temperature), the Boyle's Law and General Gas Law equations should give a general idea on how gases will behave under vacuum. Certain gases can often be pumped at different rates depending on conditions. Nitrogen is usually the easiest to pump. The percentage in a pumped down system will be dramatically lower than at atmosphere. If a leak is suspected, the amount of nitrogen present can give an indication of what is causing the problem. In a leak free vacuum, under a vacuum of about 10^{-3} torr, an analysis of the residual atmosphere generally reveals that less than 0.1% of the original atmosphere remains. The gases that usually are present consist mainly of vapors from greases and lubricants on door seals or parts, and some water vapor. The oxygen content in reference to atmospheric pressure should be in the order of 1 to 2 ppm, which is approximately equivalent to a dewpoint of –79 °C (–110 °F). At a vacuum of 10^{-4} or greater, the equivalent dewpoint is on the order of –90 °C (–130 °F). The low equivalent dewpoints achieved under a vacuum can almost guarantee contamination-free work. This is an important point to remember when troubleshooting or looking for the cause of contaminated work. A Residual Gas Analyzer will identify (a) the composition of the atmosphere in a vacuum, (b) whether a foreign substance is responsible for the contamination of the furnace, and (c) the existence of a leak. In a vacuum furnace, the amount of contamination and the vacuum level are relative to the type of work being done: that is to say, what is acceptable for one process or material may not be acceptable for another process or material.

Troubleshooting and System Calibration

Essentially, the entire vacuum system can be broken down into three main components: (1) the vessel, or working chamber where a vacuum is being created; (2) the pumping system, which produces or evacuates gases in the vessel to create a vacuum; and (3) the gages and controls, which control the entire process.

The Vessel

The main concerns associated with the inspection and calibration of the furnace chamber itself are basically associated with leaks in the chamber, and the ability of the furnace to maintain both a satisfactory vacuum level and uniformity in temperature. Table 11-3 gives examples of what to look for when inspecting the furnace chamber. Leaks and outgassing are the problems encountered most often with the furnace chamber. These problems will be responsible for both poor system performance and work of poor quality.

Leak Detection and Troubleshooting

Before troubleshooting or leak detection can take place it is necessary not only to characterize the system and determine normal operating performance capabilities, but also to become familiar with the behavior of each of the components of the system. Any deviation from the norm will be apparent. A gradual rise in the number or frequency of any deviation may indicate a problem and/or the need for routine maintenance. This is an excellent area for the application of Statistical Process Control (SPC) to monitor system performance.

A basic monthly procedure should be established to determine system performance. This procedure should be performed on both new equipment and equipment that has experienced major repair or modifications and should record: (1) the Ultimate Vacuum Level, (2) the leak-up rate, (3) the pumpdown rate, (4) temperature control, (5) power usage at temperature, and (6) power usage by the pumps—both mechanical and high vacuum.

To find the Ultimate Vacuum Level:

1) Conduct the test on an empty furnace.
2) Start with a system bake-out at a temperature which is at least 38 °C (100 °F) above normal maximum operating temperature for at least 1 hour.

Table 11-3 Vacuum System Inspection and Troubleshooting Guide

Problem	Possible Cause	Solution
	Vacuum Chamber	
Leaks		
Contaminated Discolored Parts	Contaminated Gas Supply Gas Purity	Check Gas Supply for Purity
Poor Ultimate Vacuum	Contaminated Dirty Furnace	Check Gas Supply System for Leaks
Inlet Pressure Surge	Leak in Chamber	Dirty Hot Zone-Run Clean Up Cycle
High Leak Up Rate	Improper System Operation	Use Helium Leak Detector to Find and Repair Leaks
	Improper Air Release Procedure	Check All Threaded Fittings; O-Rings; Valves; Seals
	Dirty Work-Outgassing	Check Valve Sequencing
		Do Not Release Roughing Line Through Pump Ballast
		Clean Work
Temperature Uniformity		
Poor Heat Control	Check Power Control Unit (PCU)	Repair or Replace as Necessary
	Check Element Voltage	Check Instruments and Thermocouples for Accuracy
	Check Control Instruments and Thermocouples	Clean/Repair or Replace Insulators
	Check Insulators for Damage	
Poor Temperature Uniformity	Hot Spots in Furnace	Check Cooling Jacket Water Supply for Contamination and Deposits
	Check Work Control Thermocouples for Damage and Accuracy	Calibrate Instruments
	Check Control Instruments for Drift Accuracy	Do Temperature Uniformity Survey
	Check Trim Controls	Replace Thermocouples

Problem	Possible Cause	Solution
Vacuum Chamber (continued)		
Temperature Uniformity (continued)		
Heat Failure	Check Hot Zone Fuses	Replace Fuses
	Broken Heating Elements	Repair or Replace Heating Elements
	Check Fuse in Power Control Unit (PCU)	Repair or Replace Control Units as Necessary*
	Check Variable Reactance Transformers (VRT); PCU and Temperature Controls	
Ultimate Vacuum Level		
Unsatisfactory Ultimate	Leaks in System	Detect Leaks and Repair
	Material or Work Outgassing	Clean Work or Adjust Cleaning Method to Leave No Residue
	High Forepressure	Check Mechanical Pump Performance
	Vacuum Gages and Instruments Malfunctioning	Test Gages and Instruments vs a Calibrated Standard
		Calibrate or Replace if Necessary
Pumping Systems		
Diffusion Pump Performance		
Unsatisfactory Pumpdown	Low Heat Input to the Diffusion Pump	Check Operating Temperature of Pumps
		Check Voltage to Diffusion Pump
		Check for Broken Heating Elements
		Check for Damaged Jets and Pump Assembly
	Low Oil Level in Pump	Add/Change Oil
	Contaminated Oil/Mixed Oil	

*Refer to manufacturer's instructions

Pumping Systems (continued)

Problem	Possible Cause	Solution
Roughing Pump Performance		
Unsatisfactory Pumpdown	Low Oil Level in Pump Contaminated Oil, Mixed Oil Gas Ballast Valve Opened Air Release Valve Open Backfill Gases Entering Chamber Worn Out Pump	Add Oil Change Oil Close Valves Make Sure Process Controller is Sequencing Properly Check Pump Blank-Off Pressure (Pressure When Roughing Valve is Closed)
Leak in Valve or Fitting	O-Rings Worn Out Valve not Seated Properly	Replace O-Rings Clean or Replace Valves
Instrument Calibration and Process Accuracy		
Exceeding Allowable Tolerance of Instrument	Failure of Instrument Tampering with Adjustments by Uncertified Personnel Cleanliness and Stability of Instruments Environment Component Drift	Repair or Replace with Calibrated Instrument Allow Only Authorized Persons to Make Adjustments Isolate Instruments from Large Temperature Fluctuations and Dirt Calibrate and Check for Accuracy on a Scheduled Basis

Process and Temperature Control Instruments

Problem	Possible Cause	Solution
Process and Temperature Controls		
Poor Heat Control	Furnace Dynamics Change Problem with PCU Instrument Out of Tolerance/Calibration Drift in Thermocouples	Look Beyond Instruments for Cause of System Performance Adjust Per Manufacturer's Instructions Calibrate Instruments - Certify Furnace Temperature Uniformity Replace Thermocouples
Heat Failure	Loss of Power Check Power Control Unit, VRT, and Temperature Controllers	Check Power Source Repair as Necessary*
Failure to Enter Heating or Cooling Cycle	Check Control Relays Check Solenoids Check Process Controller	Repair or Replace Broken Relays, Solenoids Check for Proper Program Cycle
Vacuum Level and Accuracy		
Unsatisfactory Vacuum Level	Check Sensors and Instrument Check Process Control Instrument Check Process Cycle Check for Leaks Check Pump Performance	Compare 1:1 Versus a Calibrated Sensor and Instrument Check for Accuracy and Calibration Status Make Sure Programmed Cycle is Correct Find and Repair as Outlined Above

*Refer to manufacturer's instructions

3) Then, turn off the power to the hot zone and and allow the system to pump overnight. When the temperature in the furnace is near 38 °C (100 °F) or less, record the pressure.
4) This is the Ultimate Vacuum Level.

The next step is to isolate the chamber by stopping the pumps, and record the leak-up rate. To find the leak-up rate:

1) Pump the system down to a pressure below 10^{-3} torr.
2) Valve-off the chamber to stop further pumping.
3) Record the rate of rise in the system pressure.

Note: Since everything leaks slightly, the pressure in the furnace chamber will rise over a period of time, but very large leak-up rates or a changing rate of leak-up can be indicative of a problem.

Outgassing is due to the release of low pressure contaminants present in the system. In order to eliminate normal outgassing:

1) A minimum of 15 min. to 1 h must be allowed for pumpdown.
2) Normal leak-up rates are in the range of 5 microns/h. If the leak-up rate seems excessively high, vent the system to a higher pressure and begin pumping down for the same amount of time, or pumpdown to the same pressure as the first time.
3) Valve the system off, and record the leak-up rate.

If the leak-up rate improves it may be an indication of excessive outgassing. This procedure should be repeated as many times as it takes to achieve an acceptable leak-up rate and until you feel confident that there are no other problems. However, if the leak-up rate does not improve or worsens, it may mean there is a real leak in the system.

After a leak-up rate is determined, backfill the furnace chamber to atmosphere and begin the pumpdown cycle and record the time it takes to achieve certain vacuum levels: such as the time from atmosphere to crossover pressure (80 to 100 microns), and the time from the crossover to any pressure 10^{-x} torr. Initial pumpdown curves can be made by plotting the time to pump to a certain pressure. Use this data as a basis for comparison between the initial condition of the furnace and the present condition. Figure 11-6 shows an example of a pumpdown curve. If subsequent curves are plotted, over time a gradual shift to the right may be observed. As long as the curves have essentially the same shape, this gradual shift is most likely due to the

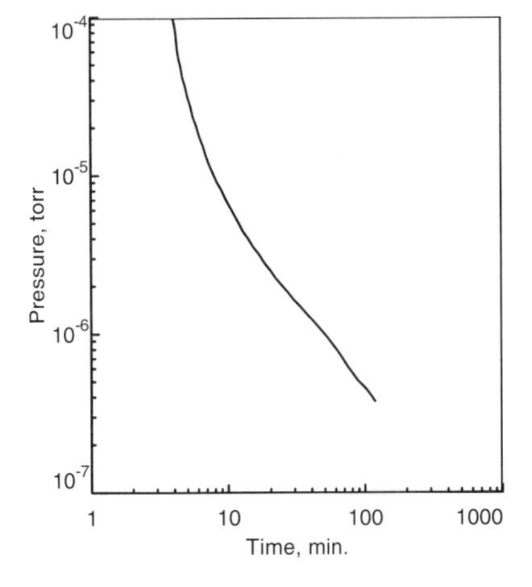

Fig 11-6 Sample system pumpdown curve

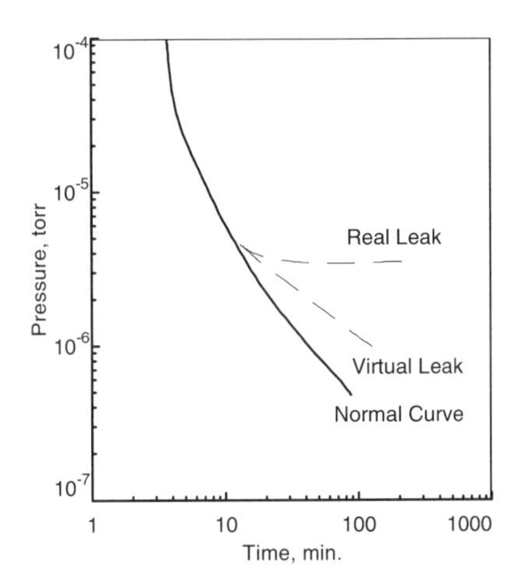

Fig 11-7 Pumpdown curves. Virtual and real leak as shown by pumpdown curves

presence of foreign material, grease, oils, *etc.* which have accumulated in the furnace over time. Routine maintenance and cleaning are indicated when the pumpdown times begin to take too long.

Pumpdown curves may also help to alert operators to the presence of a virtual or real leak (Fig 11-7). Virtual leaks are caused by the slow release of gases into the furnace from a trapped source (*i.e.*, lubricants or grease trapped inside holes in parts, or from gases trapped between O-ring seals). On the other hand, real leaks are caused by some other problem (*i.e.*, improperly sealed fittings, cracks in a weld, water leaks, *etc.*). A pumpdown curve will show the system pumping down to a certain pressure where the incoming air matches the ability of the system to expel it, and essentially the curve will become a straight line at that point.

Actual Leaks

If an actual leak in the system is suspected, an attempt must be made to determine the source: that part of the system that has the leak. If discolored parts are found, the first step is to rule out the possibility of a contaminated quenching system. In systems that draw nitrogen or some other quench gas from a main tank servicing an entire plant, discoloration of the parts closest to the quench ports in the furnace could indicate a contaminated quenching system. Usually, parts not hit directly with the gas will be unaffected. Indeed, if a contaminated quenching system is the cause, other areas that feed off the same lines in the system also will be affected. However, if it is found that the gas system is not the source of the leak, the system must be pumped down and valved off. Then systematically add parts to the system and record the leak-up rates. Usually, this method will isolate such areas as the blower or a pump. Once the area of the leak is determined, the actual leak itself must be pinpointed.

The actual leak may be a combination of many small leaks and the ease with which it is found probably will depend on the size of the leak-up rate. Helium permits the dynamic testing of a system because it is the lightest and smallest of the inert gases, and therefore can penetrate small leaks quite freely, is easily separated in a mass spectrometer, will not contaminate parts, and is able to migrate upstream through diffusion pumps. Therefore, a helium mass spectrometer is commonly used to locate the presence of leaks in a vacuum system. A soap solution can be used to bubble check for leaks for the parts of the system above atmosphere. A typical set up for a helium leak detection system is shown in Fig 11-8. Samplings are usually taken from the foreline portion of the system located at point "A."

Certain types of leaks will show up on a helium leak detector due to the presence of virtual leaks in the system, but will be extremely difficult to fix. One of the common causes of this type of virtual leak is the presence of gas trapped between a double weld on a flange or a vacuum chamber. The gas

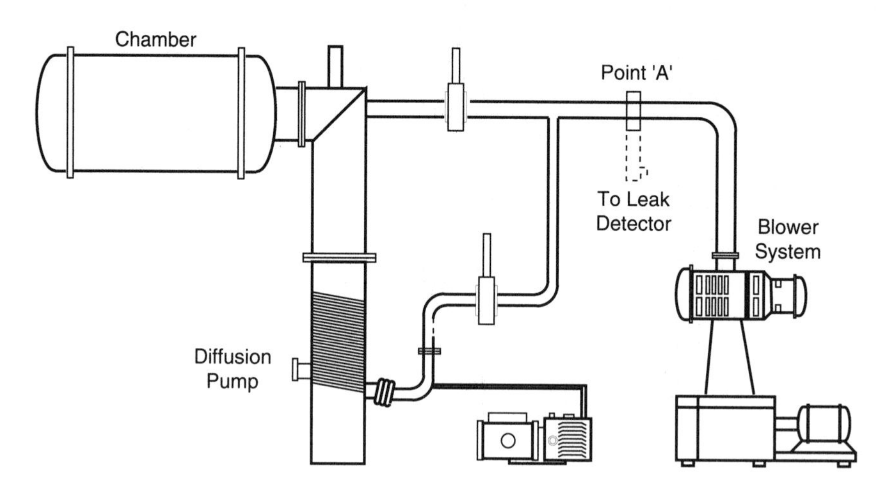

Fig 11-8 Leak detection set up

trapped here flows slowly from the weld through the inner leak into the chamber. If the leak is from the inner chamber only, application of helium to the outside wall of the chamber will not locate it. Similarly, if there is a leak on both sides of the weld there is no guarantee that both leaks will be in the same location on each side of the weld. Therefore, helium applied to one location on the outside of the weld may enter the chamber at a totally different point, giving the impression that the entire leak has been found.

Another possible but less common cause of a leak could be the use of porous materials (such as cast aluminum) in flanges and other parts subject to vacuum. This type of material can be the source of both virtual and real leaks, as welding porous material may open up a porous passage through which gases may enter the chamber. Further, even though it is not welded, this porous material still may contain trapped gases.

A water vapor leak in a heat exchanger is an extremely difficult leak to find in a vacuum furnace. The easiest way to find this leak is to thoroughly blow the water from the coil, and pressurize the coil with helium. If there is a leak it will show up as a leak on the leak detector.

Adding a Residual Gas Analyzer, or RGA, to your leak detection system is an effective approach to leak detection. RGA's can give clear information on the type of gases and elements present in the system, and can help pinpoint the cause of leaks (*i.e.*, an excessive amount of water vapor might indicate a leak in the heat exchanger or cooling jacket). Many RGA's have built-in leak detection modes for detecting helium and other gases that are only trace elements in air. This enables the RGA to serve as a process control instrument as well as a leak detector.

Temperature Uniformity Surveys

Generally, temperature uniformity surveys are performed once a year, or as often as once a month, depending on the type of work being processed and the processing specifications governing the work. The temperature recorder must be able to maintain a certain degree of tolerance over the temperature range being surveyed. Thermocouples must be calibrated and the thermocouple correction factors at the various temperatures also must be added to the survey temperatures. Both the temperature recorder and the thermocouple must be traceable to NIST. Most vacuum furnaces should be capable of maintaining ±5 to ±10 °F uniformity among the thermocouples in the survey. The range of temperatures surveyed must include the range of operating temperatures normally used, with at least 2 or 3 temperature set points being recorded. The heating curve must be plotted to show any overshoot which may occur, and a minimum of half-an-hour at each temperature should be recorded.

The inability of the furnace to maintain decent uniformity can be caused by something in the power control unit, the inability to trim power to the furnace heating elements, the design or condition of the furnace hot zone, and/or the accuracy of the temperature controller and thermocouples.

Types of Pumps Used in Vacuum Systems

Removal of the gases present in a vacuum system requires a combination of pumps, and is dependent upon the vacuum level to be achieved. Most furnace systems are designed to produce the most efficient pumping rates for the application each system is used for. The rate at which gas is pumped through a system is known as Through-put, Q, and is defined as the pumping speed in liters per second times the pressure. The Through-put, Q, will be the same at one end of a vacuum system as at the other, but may vary at different points in between. Although the speeds and pressures may vary, they will combine to give a single system Through-put. This single system Through-put is important when selecting pumps to work together.

The types of pumps used in vacuum systems are classified as to the level of vacuum that they can achieve (see Fig 11-1). For a rough vacuum, the pumps are mechanical pumps which are classified as roughing pumps. Some common pumps used to evacuate systems in this range are rotary oil sealed pumps, roots pumps, sorption pumps, oil booster pumps, and rotary vane or piston pumps. All of these pumps rely on the mechanical motion of their parts to trap and force gases out of the system. Depending on the type of pump and the conditions, these mechanical roughing pumps can produce vacuum levels in the 1 (10^{-3} torr) to 50 micron range. The pumping ability of mechanical roughing pumps will be greatly affected by the presence of water vapor in the air being pumped, and since water has a high vapor pressure in relation to the oil, the ultimate pressure obtainable will be decreased. At room temperature water vapor will migrate from the system at approximately 20 torr. To overcome water vapor in the pump the oil must be kept hot enough to keep condensation to a minimum, but not so hot as to reduce the viscosity of the oil. Gas ballasting—the addition of dry air into the pump near the exhaust valve—will reduce the partial pressure of the water vapor and prevent condensation in the pump.

During a pumpdown the roughing pumps remove approximately 99% of the air in the chamber. For pressures lower than 1 micron another type of pump must be used to evacuate the system further, and the two most common types of pumps are the oil diffusion pump and the cryogenic pump. These pumps are capable of producing vacuums in the 10^{-4} range and lower. Diffusion pumps operate by heating the oil and forcing the vaporized oil upward through a chimney to a jet assembly near the inlet or top of the pump (see Chapter 1, Fig 1-3). As the vaporized oil reaches the jet assembly it is ejected downward and outward together with any gas molecules it may have encountered. The oil and gas molecules are condensed, compressed, and exhausted at the foreline, and then through the mechanical pumps. To stop any gas that is trapped in the oil from being re-released above the pump and into the system, the foreline is kept hot in order to release any gases and exhaust them to the mechanical pump. Further, to condense the molecules of oil and gas and to eliminate any backstreaming, a cold trap is usually located above the diffusion pump. As the temperature of the cooling medium of the trap is decreased, the ability of the trap to serve as a pump for condensable vapors is increased. Cryogenic pumps use liquid nitrogen at around –196 °C to condense and pump large volumes of water vapor and condensable gases to about 10^{-8}.

Pressures in the range of 10^{-8} and lower can be produced with ultrahigh vacuum pumps, such as ion pumps and turbomolecular pumps. Generally, these pumps are found in laboratory or research environments. For most heat treating applications the pressure achieved with a combination of roughing and high vacuum pumps is suitable.

Whatever combination of pumps is used to achieve the required vacuum levels it is important that the operating instructions are followed, and that routine maintenance is performed on a regularly scheduled basis. The following should be avoided during the operation of a mechanical pumping system, and should be looked into very thoroughly when troubleshooting such a system:

- Dirt or contamination in the oil
- Lack of lubrication
- Too high of operation temperatures
- Broken parts
- Water, solvents, or noncompatible oils in the pump
- Ignoring routine maintenance of filters, seals, and oil
- Pump at atmospheric pressure for extended periods of time

All of the above situations could lead to eventual pump failure.

Similar rules apply to the operation of diffusion pumps. In general the following should be avoided:

- Letting atmospheric air into the pump
- Operating with low fluid levels
- Operating the pump without cooling water
- Draining or filling the pump while under vacuum, or while hot (150 °F or greater)

Always follow the manufacturer's guidelines and instructions for all pumps.

Vacuum Measurement

Some gages are capable of measuring a vacuum to only a certain level (Fig 11-9). Basically, gages fall into two categories: (1) those which measure the pressure hydrostatically, and (2) those which measure a certain property of gas that is relative to its pressure in the system. Often it is necessary to use a combination of gages in order to cover the entire pressure range for a particular system.

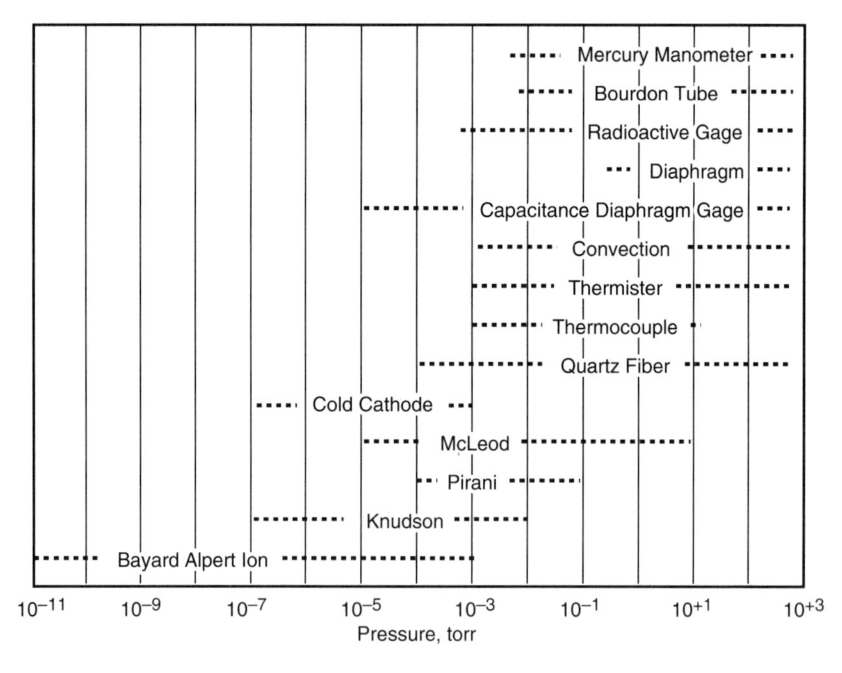

Fig 11-9 Pressure ranges of vacuum gages

McLeod Gage

Used mainly to calibrate other gages, this gage is based on volume. Comprised of a series of glass capillary tubes with a reservoir of mercury, its calibration is unaffected by gases in the system. As long as all condensible vapors are trapped out, accurate readings can be obtained and the calibration of the instrument will apply to all gases obeying Boyle's Law. On entering the gage from the system, a sample of gas is compressed to a calibrated volume in the capillaries, and the gage will register the ratio of the initial volume of the gas against its final volume. The ratio is given in terms of pressure and is indicative of the pressure of the system when the sample was taken. For this reason it is not a continuous reading gage, and it cannot indicate sudden changes in pressure or outgassing. This gage is also incapable of compensating for any condensible vapors present in the system. More accurate readings will be obtained when a cold trap is used in conjunction with the McLeod gage.

Bourdon Gage

This is a simple and reliable gage for measuring pressure in inches of vacuum. This gage is constructed so that one end is open to the vacuum chamber and the other end is sealed and connected to an indicating needle. As the pressure decreases, the difference between the inside of the tube and the outside of the tube—which is at atmosphere—causes the curvature of the needle to change. This change in curvature reads pressure directly onto a calibrated dial indicator. As the accuracy of this gage is not very high, the gage is found more generally on vacuum chamber doors. It is used as a quick reference for pumpdown or pressurizing the chamber when readings up to one atmosphere are useful.

Thermal Conductivity Gages

Thermal conductivity is the ability of the gas in the system to conduct heat away from a hot filament. The Pirani gage and the Thermocouple are the two general types of thermal conductivity gages and both are based on the principle that the thermal conductivity changes with pressure. These gages have the ability to conduct heat changes as the pressure changes, and generally are accurate in the range of 1 to 100 microns. A constant voltage and current are applied to a filament which in turn heats up. As the volume of gas in the chamber is pumped out, the number of molecules of gas that conduct heat away from the filament decreases, and the filament becomes hotter. When there are too few molecules to conduct the heat away, the filament will eventually reach a maximum temperature. The temperature of the filament can then be measured and converted to a pressure.

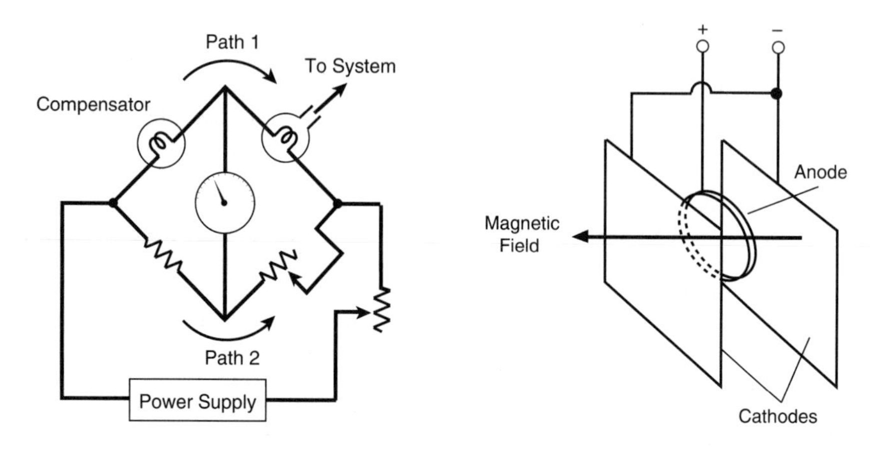

Fig 11-10 Pirani Gage

Fig 11-11 Cold cathode gage

The difference between the Pirani gage and the Thermocouple lies in the method by which the temperature is measured and then converted to pressure. In a Pirani gage (Fig 11-10) as the temperature increases the resistance of the wire increases, changing the voltage and current in a balance whetstone bridge type circuit. This current is measured and calibrated in pressure units versus the McLeod gage at a given pressure.

In a Thermocouple gage, a direct temperature reading in millivolts is taken from the hot platinum filament and is calibrated directly into microns.

High Vacuum Gages

These gages also can be classified as ionization gages. The cold cathode ionization gage is found on most furnace systems, while the less rugged hot cathode ionization gage is intended primarily for laboratory use. Both gages make use of free electrons travelling in long helical paths around a positively charged grid hoping that the electrons will collide with the gas molecules and ionize them. Then, the ionized gas molecules flow to a collector and produce an ion current proportional to the number of molecules, or pressure, in the vacuum chamber.

Cold Cathode Gage

The cold cathode gage is used primarily in industrial applications because of its ruggedness, as its elements are not directly exposed to the vacuum chamber and are not as susceptible to contamination as is the hot cathode gage. The reliable pressure range of this gage is in the range of 10^{-3} torr to 10^{-7} torr with the upper and lower limits being determined by the composition of the residual gas (cleanliness), and the amount of ion current that can be practically measured. Figure 11-11 shows a schematic of the cold cathode gage. Simply stated, the gage has a central anode and two cathodes which emit the electrons, surrounded by a permanent magnet. The free electrons which are emitted are driven toward the positively charged anode in a long helical path forced by the magnet. The long path increases the chance of a collision with a gas molecule creating a positively charge ion. These ions then travel directly toward the negatively charged cathodes and the ionization current that is produced can be directly read on a micro-ammeter in terms of pressure.

Calibration and Certification of Vacuum Instruments and Systems

The demands being placed on heat treaters are becoming more stringent each year as companies and consumers become more quality oriented. There are numerous quality standards governing the quality and integrity of the instruments and the systems, as well as the overall management and quality program of the company itself. However, at this time there is no one overall governing standard for the users of vacuum furnaces. Depending on the type of work being done, two standards apply: AMS 2750 and MIL-STD-45662A. These cover the accuracy and traceability of the thermal processing and measurement equipment, and vacuum measurement and recording devices. MIL-STD-45662A contains requirements for establishing and maintaining a calibration system for the instruments used in testing and measurement. AMS 2750 specifies the requirements for system accuracy and calibration frequency for pyrometric equipment. The frequency for periodic tests and calibrations are based on the type of work being done, and the history and reliability of the equipment being used.

All calibration system requirements (for thermocouples, hardness testing blocks, vacuum gages, *etc.*) call for traceability to a national standard. As vacuum furnaces are being used for heat treating, brazing, and other thermal processes, calibration of the vacuum gages, sensors, temperature controllers, and the furnace chamber must be done on a periodic basis. Calibration may be required as often as once a month, or perhaps once every six months, and will depend entirely on the work being processed.

Calibration of Vacuum Measurement Equipment

Most vacuum furnace systems can be calibrated with the connection of a certified master gage in parallel to the existing instruments of the furnace for a one-on-one comparison. Calibration should be done during a cold pumpdown of the system, and is best done after a clean-up cycle has been run as this will ensure that the furnace is free from contaminants. Annual recalibration of the master gage tubes and cables must be done by a calibration laboratory with standards traceable to NIST. NIST certification can be applied only to a given instrument, gage, and cable. Various types of gages are used as standards and can be kept as inhouse standard gages. Some of the primary standard gages used for calibration are: McLeod Gage, Spinning Rotor Gage, Hot Cathode Ionization Gage, and the Capacitance Diaphragm Gage.

Most vacuum gages should be calibrated at least once a year, but certainly more often if the system does not seem to be performing satisfactorily, and/or the work requires that the system be calibrated more often.

Conclusion

The basics for the inspection of vacuum systems have been discussed in this chapter, together with general guidelines for the calibration and certification of a system. However, the successful use of any equipment requires that the manufacturer's operating and maintenance instructions must be followed implicitly. Also, there is no substitute for well-trained employees. As they are involved in the day-to-day operation and maintenance of the vacuum furnace, this training will make the difference between an operator who catches problems and the operator who causes and/or ignores potential problems.

References

1) Abar Ipsen Technical and Training Data

2) L. Van Roekel and W.C. Worthington, "Gas Analysis and Leak Detection for Vacuum Heat Treating," Leybold Inficon Inc., Technical Manual, 1992

3) W.H. Bayles, Jr., "Vacuum Measurement, Calibration, and Certification," Paper presented at the First Annual Vacuum Users' Conference sponsored by Abar Ipsen IND., October 1992

4) D.B. Webb, Leybold Vacuum Products Inc., *Heat Treating*, Vol 24, No. 1, January 1992

5) "Much Ado About Nothing, Or ... So You Want to Measure Vacuum?" The Fredericks Company

Index

B

C

G

H

I

K-L

N

Q

R

S

T

U

X-Y-Z